BACKYARD DAIRY GOATS

A NATURAL APPROACH TO KEEPING GOATS IN ANY YARD

For Geraldine, Felicity, Wrinkle, Pin, Moon, Pixie, Snowy, Ned, Leaf, Fern, Grass, Liflin, Sunshine, Fehu, Ostara and all the other goats.

I would like to thank everyone who has helped make this book possible. To each and every one of my Kickstarter backers, and everyone who has helped spread the word of the book – I thank you all for helping create something that is in some ways a truly independent book, and in other ways a real community effort. It is like planting a food forest or building a goat barn together; we have done this by all helping when the help was needed and created something great that I hope will benefit goats and those that love goats.

There are families, farms, and individuals who have been especially supportive of this book, in particular I would like to name Nicole of Fairy Grove Creations, Julia Winter, Matthew, Cécile, & Mary, Jay Skiles, Goat Berry Farms, John Weaver, Joe Salcedo, Fertile Roots Foundation, Sarsbar, Garry Platt-Hepworth, Llovet Family, Joseph Alexander Chugg, Richard Wilkin, Brandi, Kat Fryzb, Melissa J. Edwards, Thadius Marcus, Rosemary Schmidt, Permaculture Kindergartener, Laureen Sue Magyari, J. Lauryl Jennings, Courtney Siobhan, Foliage, and Wendy The Weird Plant Lady O'Neill. I couldn't have done this without every one of you. Thank you!

An extra special thank you with bacon, goat cheese, sparkles and pie goes to the community at Permies.com. The team at Pip Australian Permaculture Magazine deserve a big thank you for their support and generosity. I would also like to thank Permaculture News, and the Australian Raw Milk Movement.

A big thank you to my husband for his help, encouragement, and occasional sense of humour. Thank you also to my lovely children.

And to those who are not named here (you know who you are) -
Thank you!

CONTENTS

INTRODUCTION

UNDERSTANDING GOATS

THE NEEDS OF BACKYARD GOATS

GETTING YOUR GOATS

DAY TO DAY GOAT KEEPING

CHEESEMAKING AND RECIPES

INDEX

A morning with goats

You might start your day with a glass of cold raw milk from yesterday's milking, or maybe hot tea with milk. You might eat homemade yoghurt for breakfast, or some toast with fresh chévre. Opening the door to refill your goats' food and water, you hear their bleats. They do this often when they hear the door close, looking forward to gobbling up broccoli stems, outer lettuce leaves, carrot peelings, and other treats you've brought them. They wag their little tails and lift their heads over the fence for a pat as soon as they see you. Goats bring so much joy to your day, it would be worth keeping them for pets even if they didn't provide you with nourishing raw milk and all the cheese you could possibly eat. Goats can be happily kept in a backyard, contributing to your kitchen, garden and family life. This book will show you how.

My story

Goats are one of my favourite animals. I have always appreciated the cute way that they nibble the leaves off trees, the way they bleat, their goatish personalities, and just the general appearance of a couple of goats going about their lives, which for most of my life I had only encountered from the other side of the fence.

For many years I have aspired towards self-sufficiency and permaculture. In practise over the years I kept vegetable gardens in different rental properties, never able to really appreciate the beneficial interactions that animals could have had with these gardens.

It wasn't until I learned of the benefits of raw milk that I put these two interests together. Our family moved from a forest suburb near a city to a cottage on a third of an acre in the country, our goal was to save up for some land of our own, but what I learned in this time moved us towards greater self sufficiency well before we were able to own land.

It was in this time that I learned the benefits of traditional foods. Here in Australia (at the time of writing) selling raw milk is illegal, and as a result it can be very difficult to get hold of some. When we did manage to buy raw milk, the supply was irregular, and it meant extra trips out in the car when we would rather stay at home.

I began to look into the possibility of keeping a couple of dairy goats in our yard. This meant asking a lot of questions, and some of them remained unanswered, or had very vague answers. Existing books about goats were written by people who kept them on a much larger scale. Most of them were filled with stuff about showing goats at goat shows, weighing the milk to keep records, dehorning goats and other things that I didn't need to learn about, while not answering the questions that are most important to someone on a budget wanting to raise goats for milk: How much space do they need? What food do they eat and how much of it do they need? How do I keep them naturally healthy? How do I sterilise milking stuff without chemicals? What do I need to keep goats in a backyard?

A lot of these things I had to learn as I went. We bought a transportable fence designed for dogs, a moveable shelter, and not long afterwards we found two pregnant Toggenburg does for sale nearby. They had been bottle-raised as pets. The family had milked them a couple of times for fun but couldn't answer many questions about it, and the goats had weaned their last kids and weren't in milk anymore. I didn't know much about goats back then, but I liked these ones, they were a pure dairy breed, and the people selling them seemed honest, so we brought them home and looked after them for the rest of their pregnancies, mainly following Pat Coleby's "Natural Goat Care" for advice, and applying it to a much smaller scale situation.

One morning my husband came to tell me that one of the goats had made a strange bleat, and he thought she might be giving birth. I went outside to look and found her licking a healthy baby goat, and observed as she and her baby followed their natural instincts to bring him to his feet, and to her udder. Not long afterwards the other doe seemed to be behaving differently to usual, she went into the shelter and paced around restlessly, fluffing up the straw in the hut to form a nest. I observed the entire birth of her twins.

A few days after this, I milked a goat for the first time. All the preparation we'd had over the past couple of months, leading the goats up to the milking stand and feeding treats while I patted them had been worth it, and milking wasn't much of a learning curve for them or for me.

The kids stayed with their mothers until around three months of age, when I found new homes for them. After this time I was milking twice a day and began to experiment with cheesemaking. I had some good books for this, but a lot of it was still trial and error, and I felt uneasy about using these strange powdered cultures in the recipes and eventually began to make cheese using only natural cultures that I could keep alive indefinitely at home.

My goats stayed in milk for three years, one of them having a break from it due to an udder injury, and then coming back into milk on her own the next spring. We moved house a couple of times during these years, eventually finding our way to a place of our own. Another goat joined our herd, then the opportunity arose to have a buck on our land for a few months. Our herd grew, and shrunk, and grew again. I have had time for many observations and cheesemaking adventures, and continue to learn as I observe and interact with the goats each day.

Every day I seem to appreciate more the simple beauty of the natural mixed farm, where animals feed the soil, the soil feeds the plants, and the plants then feed the animals.

I have a website at thenourishinghearthfire.com, where I will post any new updates to this book as new things are discovered about goats.

UNDERSTANDING
GOATS

A goat's place in the backyard ecosystem

Raw milk for health, survival, and economy
There is nothing like having a reliable year-round supply of fresh milk for the family. Raw milk is a complete food, rich in protein, vitamins, and minerals. Having a source of this will reduce your household's need for other animal proteins, and will give reassurance if there are picky eaters or vegetarians in the family who might be missing out on nutrients from other foods.

With a couple of goats for milk, a few hens for eggs, and a vegetable garden it's possible to produce all the calories and protein you need in a typical backyard. Goats milk can be turned into delicious cheeses that can become a staple part of the diet for much of the year.

While many people think of preparedness in terms of having dry food stored away, having raw milk as a source of protein, fat and calories that appears fresh every day without needing much food to produce it is something that can make other preparedness efforts easier and healthier.

If you're used to buying milk in plastic bottles, keeping dairy goats will also greatly reduce the amount of plastic waste your family produces. If you buy milk in returnable glass bottles, keeping goats will save you money. If you go out on lots of trips to the shops or farmer in order to get fresh milk, keeping goats will mean less time in the car.

Optional goat meat
While it is possible for vegetarians to milk goats without the goat getting in kid every year, for those of us who do eat meat, the offspring of our dairy goats can be a good source of red meat and bones for broth.

Whey for pigs
Pigs thrive on excess milk and whey (along with foraged acorns, windfall apples, and other often-wasted foods), so raising a couple

of pigs when you have an abundance of these provides more self reliance and a greater variety of nourishing foods to the family diet.

Pets and lifestyle

Goats are intelligent animals that have lots of personality. It is a joy to spend time with them, watching them as they eat broccoli stalks, have play fights, and climb on cable spools. The routine of daily milking makes us more appreciative of our goats, what they are providing, and allows time for observation and reflection outdoors.

Compost and manure generators

Goats form a synergistic relationship with the backyard veggie garden and orchard, providing manure and bedding to compost or use as mulch. Goat manure doesn't stink, and is a fairly mild manure that can be used on plants right away. Excess goat milk and whey can also be used on the garden.

Waste disposal

Goats can be fed many kinds of garden scraps and tree prunings, converting them into milk for the home, and manure for the garden.

Land clearing

If you have blackberries invading part of your land or common land nearby, goats will gladly gobble up the invasive plants, thriving on the minerals in them.

A note on goat Newspeak

Like much of the language and etiquette in the modern world, the correct way to refer to goats is often confusing. Some people will get terribly offended when you refer to a male goat as a billy, others won't know what you're talking about if you say 'buck' and 'doe' instead of 'billy' and 'nanny'. I will translate the goat Newspeak for you.

The correct term for a mature female goat is a doe, a female goat kid is a doeling. If you're searching for goats on internet classifieds such as Gumtree and Craigslist, it's worth checking for 'nanny', or just 'goat' as well as 'doe', as there might be some good milking goats on there that are not listed as does or doelings.

The correct term for a mature male goat with his balls intact is a buck. The term for a male goat without balls is a wether. Bucks can be difficult to deal with, and their meat tastes bad, so most goat owners will wether their male goat kids when they're young. If you're searching online for a male goat for breeding, it can be worthwhile to search for 'billy' or just 'goat' instead of buck.

Goat behaviour

Herd dynamics

Goats are herd animals. While sometimes there is a rare goat that is happy without other goats, the vast majority are miserable without another goat as a companion. If you don't think you need much milk, it might make sense to have a wether as a companion for your doe. If it's important to you to have milk every day of the year, or you want more milk than one goat will provide, then it makes more sense to have two does and breed them in alternate years, or breed one early in the season and the other late in the season, so that at least one of them is in milk at any one time.

Goats have a strict hierarchy where the head goat (also known as the herd queen, or alpha goat) needs to be the one to get the first share of the food, to be milked first, to get the first pat and so on. If you approach another goat with veggie scraps, the head goat will usually butt the other goat out of the way to get to the treats first. Goats will playfully fight from time to time, and the head goat will always win this fight. She's usually the larger of the goats, or the one with the biggest horns.

Handling goats

From the wildest goats, to someone's pets, getting to know any goat is usually a matter of bringing them lots of their favourite foods and plenty of patience. You can start by placing a bowl of food fairly close to you, and standing still as the goat eats it. Every day you can place the bowl a bit closer, eventually you will be able to hold the bowl while the goat eats from it. After a few times of holding the bowl while she eats, you can begin to pat her with your other hand, and then hold her by the collar or horn and lead her around while moving the bowl along, so that she thinks she is just moving with the food, but she is getting used to trusting you. Goats usually prefer to be gently patted on the side of the neck and under their chin rather than on top of the head. It can be a slow process, and any stressed or angry behaviour on your part is likely to slow down the process, so it is generally worth getting a goat that doesn't need to be handled right away (i.e. one that is not in milk, and not about to kid).

In the wild, goats would see humans as predators. Their instinct is to run away, and any aggressive behaviour on your part will make them think you are a predator, and lead to further frustrations. Always be calm around goats.

Ideally, you want your goats to think of you as the head goat. To help this, when you're leading them around, or holding a leg while trimming hooves, always be very determined, and never let her for a minute think that she can overpower your grip, or she is likely to continue doing this. If she realises that she can't overpower you, and that it's pointless to try and get out of your grip, then she will stop trying and you will have a much more pleasant time when you need to lead her around or hold her still.

If you don't mind so much if she does go wandering off on her own, then it's better to lead her correctly for a short time, and then let her run on her own with you deliberately letting go at a quiet moment, rather than having a battle of wills with her and then eventually letting her get her own way.

How to get a goat to appreciate the milking stand

It's a good idea to get your goats used to the milking stand as soon as possible. You'll probably need to secure them for hoof trimming every couple of months (see page 46); feeding them their concentrated food individually means that they all get their share (if you feed it in bowls in their paddock, the head goat will often finish hers first and then steal other goats' food) and when it's time to start milking them, they will associate the milking stand with treats, and will make life a lot easily by happily jumping up and eating their food while you milk them.

To begin with the milking stand, it's best for your goats to be comfortable around your presence before you try to get them to do anything, don't try to get them on the stand on their first day home unless you absolutely have to. Visit your goats regularly, talk to them, bring them treats, pat them if they let you. You can then begin holding a bowl of feed in one hand and leading the head goat by the collar with the other hand to get them used to being led around the paddock. To lead a goat, always be on guard and not distracted by other things, and hold the collar very firmly. It's best to lead for a short time to begin with, and work your way up to leading her for longer times. She may try to get away, but once she's used to being led and knows you're leading her to food she will usually cooperate. Once they're showing an interest in you when they see you, and can be led easily, you may want to try taking one to the milking stand.

To get a goat used to the milking stand, begin on a day when you are not in a rush, carefully get one goat out of the paddock, if you're not worried about her escaping to somewhere dangerous or destroying your garden, you could lure her up to the stand by shaking a pan of grain or other treats and letting her eat some every so often, alternatively, you can lead her to the stand by her collar or horns (it's easier to do this with a pan of grain to encourage her to begin with).

When she's at the milking stand, put the pan of food on the base of the stand, let her eat a little, and then hold the pan up higher so that she has to jump onto the stand to get to it. Place the feed in the feed basket on the other side of the headgate and let her enjoy her feed in peace. You can give her a pat now if you want; just observe her and make sure she's comfortable with you. Don't grab her udder straight away, just let her get used to the stand. If she's a bit nervous you may want to leave the headgate alone for now, and get her used to it later, once she's comfortable with the stand. If your goats are just one gate away from the stand, another option is to leave treats on it, open the gate, and allow them to explore it themselves, at some point they will need to get used to you being around when they're on the stand, so it's good to begin doing this as soon as you can.

It's easier to bring the herd queen out first, and take the beta goat later. You may need to have someone on hand to help get the beta goat back into the paddock when she's finished, as the head goat may lurk around the entryway, butting the beta whenever she tries to get in, and your beta goat will be reluctant to return if she knows she's going to get butted. Another option is to set up a tether for the alpha goat before you take her out of the paddock, and when she's finished on the milking stand you can leave her happily browsing while you take the beta goat to the stand.

Goats are not as predictable as they would sometimes like you to believe. You may get to the point where your goats are in a routine, happily running up to the milking stand by themselves every day while ignoring your garden, but one day a goat might change her mind and suddenly decide that she'd rather trample and eat your cabbage patch. It's best to be prepared for this, either by designing your milking area and goat enclosure in a way that they can't escape to anywhere dangerous or easily damaged if they do wander, or by keeping your hand securely on her horn or collar until she's on the stand.

Introducing a new goat to the herd
Depending on how the new goat has been raised, she will either settle into the herd quickly, or there can sometimes be prolonged bullying.

Goats have a way of establishing hierarchy, and if the new goat is not used to this behavior she may not know how to find her place in the order. As a precautionary measure, it's best to establish a separate place for the new goat, where she can interact with the other goats, but they can't reach each other. This can be a tether on the other side of the fence, or a separate paddock. If you have a buck, you can rub a rag all over him, and then rub the rag on the new goat, so that the other does don't think she is a threat to their order.

If you have the space, try to introduce the new goat to the existing herd on a different part of the yard to their usual home, that way they will be less territorial.

When you introduce the new goat to the herd, make sure to do it at a time when you will be around to observe for at least an hour. You don't have to watch them for the entire hour, but you need to keep an eye out to make sure no one is being injured or excessively picked on. It's a good idea to have someone else around to help at this time, in case you need to separate the goats. If one goat is picking on the new goat, and the others are fine, remove the bully for a while and let the new goat settle in with the other goats, before introducing the bully at another time when you are around to observe them.

When you're choosing a new goat, it's best to find one that's from a similar background to your existing herd. If your goats are from a large commercial dairy herd of docile does, then you can get away with introducing another goat from a similar background. If your goats have been living as a small herd with a distinct butting order and given a lot of freedom to express their goat behavior, then a goat from a commercial dairy may have trouble adjusting to their hierarchy, and rather than challenging the other goats when they challenge her, she may try to ignore them. This can make the other goats more aggressive, as she isn't properly submitting to them, and she isn't letting them know her place in the order by butting them back.

Always observe the goats for a while. If the new goat is trying to sit down or eat and other goats aren't letting her do this, help her by either removing the bully (or bullies), or moving her away, so that there is a fence between the new goat and the herd.

If you need to keep the new goat separately from the others to begin with, you may find one day that she can be introduced into the herd with no problems later on. One of our does, Snowy, was from a commercial dairy herd and didn't understand our herd dynamics to begin with. After a few months of living outside their paddock, we put them all in a paddock together again and she had learned their herd behaviour, and now she even asserts her place in the herd with smaller does.

Dealing with escaped goats, and herding goats

Food will always be helpful in this situation, either by calling your goats over to you while holding a familiar bucket, or putting the bucket on the ground and trying to convince the goat to go to it.

If you need to herd a goat around from a distance, it helps to know of her 'pressure zones'. She has a blind spot directly behind her, and roughly in the same wedge shape that her body follows. Goats don't like being approached from this blind spot. The pressure zones are on either side of the wedge, where she can just see you, but you are still almost behind her and not too far out to the side. If she starts to move too far in one direction, go to the pressure zone on the opposite side of her (or hold out your arm in that direction, making a swooping motion with your arm towards her) and calmly walk towards her there. She will generally keep pace with you, keeping roughly the same distance all the time, so if you run, she will run too. If you need to move her straight ahead, just keep making gentle swooping motions with each arm as she moves from one side to another, while walking calmly. The more agitated you become, the more frightened and unpredictable she will become, so try to be calm, and just to steer her away from anywhere dangerous. If she is very reluctant to go where you need her to go, leave a bucket of food on the ground somewhere safe, and be ready to either try to firmly and calmly get her by the collar or horn while her head is in the bucket, or gently move the bucket along so that she follows it.

Goat health and observation

Just as things can go wrong with humans, things can also go wrong with goats. There are ways to prevent this happening as much as possible, but sometimes things do go wrong. This chapter is more about how to prevent things going wrong, and is worth reading before you get goats. The symptom checker section in the 'Day to Day Goat Keeping' chapter is designed to be a quick reference to check if something does go wrong, not something to be read before you get goats. When goats are kept in a way that is respectful of their nature as browsing animals, and not expected to be lawnmowers, problems are very rare.

It helps to know first of all how a healthy goat looks and behaves. Observe your goats a lot when they are well, and you will easily be able to pick up if they have any health issues. It helps to observe them in different weather, as they can sometimes appear miserable when the weather is cold or rainy, but if they are behaving in good weather as they do in cold then there may be a problem.

Observe whether your goats bleat to greet you or not. Some goats do and others don't, but if one of them always greets you, and one day she doesn't, then it can be a cue to carefully observe her other behaviour, and to have a look at her for signs of problems.

A healthy goat will be curious and interested in food. She will sometimes be standing, sometimes sitting. Often when she's sitting she will be ruminating (chewing). If you observe closely when she's ruminating, you might notice a reverse-swallowing action in her throat as food is brought up from the rumen into her mouth, followed by bulges of food in her cheeks that she will chew. She will also ruminate sometimes when standing up. If she is used to you and your herd of goats, and isn't too wild, she will usually appear relaxed in temperament.

Her eye membranes when the lower one is gently pulled down as far as it will go should be deep pink or red, her coat shiny and the right colour for her breed. Her poo should come out in little dark round pellets. If the weather is good, her tail will usually be up or sticking out, if it's cold, it will usually be tucked in. She might play fight from time to time with other goats. When fighting she will go along with the other goats fighting; if she is being picked on, and not butting back then there may be a problem.

If she is being regularly milked, her udder will be a pleasant baby-bath temperature, it should not feel cold, and it shouldn't feel hot either.

If the goats all seem cold and shivering, they are probably just cold, and would appreciate a bucket of warm water to drink, but if one goat is shivering and the others are not, then there may be a problem. Observing the positions of the tails can also be a clue, if all of the goats have their tails in one position, and one goat has hers in a different one, this can mean that she is unwell. Often these first subtle signs are ones that mean that you can quickly nip a problem in the bud before it gets worse, by looking at her food and mineral intake and giving her some extra supplements if she needs them. General remedies that help many ailments, such as giving vitamin c, dolomite, diatomaceous earth, garlic, extra kelp, and apple cider vinegar will help your goat deal with any unknown illness at this early stage.

There is no need to be stressed about making these observations, they just will happen gradually as you spend time with your goats, and at some stage you will instantly be able to tell when something is not right. Just as you would get to know a person, you can get to know a particular goat and her place in the herd.

Preventing mastitis
and other udder problems

Mastitis can be caused by bacteria getting into the teat canal before it has naturally sealed itself off after milking. Observe your goats after milking, if they are always standing up for at least ten minutes after milking, then you don't have to worry about this. If they sometimes sit down, then there is the risk that they will sit on dirty ground and bacteria can enter the teat. If your goat has a tendency to sit down after milking, or if you wish to be extra careful, there are various options:

- The first is to leave her on the milking stand for ten minutes after milking. You can give her some hay or extra treat feed while she's there.
- The second option is to have an area that they're only allowed in straight after milking, this will reduce the chances of there being manure all over the ground.
- Another option is to delay putting food in her feeder until after milking time, she will then be distracted by the food and will forget about wanting to sit down.
- Post-milking teat dips might also be an option, but using these means you should be more careful about cleaning your goat's teats before milking, to make sure that your milk doesn't taste like tea tree oil or whatever else is in the teat dip.

Supplementing your goat's 'treat' feed with dolomite (or by offering it free-choice if she eats it that way) will help her to avoid mastitis. If you suspect that your goat might have mastitis, vitamin C, along with dolomite can help.

Avocado leaves can cause mastitis if eaten, so should be avoided.

If the skin on the teats or udder appears to be having problems, rub some raw goat milk on it after milking to help it heal. Continue doing this every time as a preventative measure, it will moisturise the skin to prevent future skin problems. Raw honey is also good for healing broken skin.

Preventing milk fever

High-producing dairy goats are more prone to this than any other goat. To prevent this, make sure that your goats are getting plenty of kelp and dolomite as they approach kidding, and also possibly vitamins A and D.

Preventing parasites

Read my chapter about feeding goats: learning from wild goats (page 38). If you follow the principles of this by keeping the goat's food off the ground (or doing rotational grazing), supplementing any minerals that are deficient in her feed (especially copper), then you will have a resilient goat that isn't affected by internal parasites.

For lice and other external parasites, check the base of your goats' hairs near the spine in several places, along with the legs, hips, and shoulders, look closely, and if you observe any tiny insects, sprinkle diatomaceous earth over the spine and rub it into your goats fur as best you can wherever you've seen the insects. The two days before the full moon are the most effective time for doing this. Parasites are more active in spring, so it's best to check once a week, more often if there seem to be a lot of them.

Goats that have the right amount of sulphur in their feed are less likely to have problems with external parasites. One spring, some of my goats seemed to constantly have ticks and lice, but as soon as I began adding extra yellow sulphur to their feed the external parasites went away.

Goats can generally handle a small amount of parasites if they are healthy, but when they're under physical strain, such as the next few days after giving birth, they may become more susceptible to infestation. You can help them bounce back during these times by keeping a close eye on them and giving them plenty of access to kelp, dolomite, 'treat' feed and other rich sources of minerals, as well as encouraging them to eat their placenta by leaving them undisturbed during and after birth, if possible. Sometimes the placenta will lie on the ground for an hour or two before she eats it, so avoid covering it up until after this time has passed.

To help make your goat more resilient to worms as she approaches kidding time, it may help to give her a teaspoon of copper sulphate, a teaspoon of dolomite, and a teaspoon of vitamin C powder down her throat, in addition to her usual doses of minerals.

If you're worried that your goat may have worms, try dusting diatomaceous earth on her treat feed, sprinkling it as if it were salt. Some shops that sell diatomaceous earth recommend three tablespoons as a daily or twice daily dose for an adult goat, and to continue this dose for three weeks in order to kill all the worms, but it is often difficult to feed a goat this much at once! You can try sprinkling it through her hay or chaff if you are feeding a lot of that, but it is trickier to observe whether she is eating it all or not.

Another way of feeding diatomaceous earth is to hold her mouth open with one hand while pouring a small handful of diatomaceous earth down her throat with your other hand (a film canister is handy for this job). Hold her mouth closed for a short time, to make sure she doesn't get it out of her mouth straight away, some of it will probably end up on the ground, but plenty will have stuck to her tongue and the inside of her mouth, and will be absorbed. If you are very worried about her condition, continue to give her diatomaceous earth in this way two or three times a day. It may be worth giving your goats a monthly or twice monthly dose of diatomaceous earth all through the spring and summer when parasites are most active as a preventative measure. Feeding diatomaceous earth every day during this season will reduce the amount of flies hanging around.

There are some arguments against the daily feeding of diatomaceous earth, and if you wish to try a single dose of it instead, the most effective time to do this is two days before the full moon.

Goats that are supplemented with minerals missing from the soil they're fed from, especially copper, are far less parasite prone than goats fed on mineral deficient feed. Always check first to see whether copper is in abundance in your local soils before supplementing to ensure that you don't end up giving your goats too much copper. If your soil is deficient in copper, a teaspoon of copper sulphate added to a Saanen's treat feed (or directly down her throat, mixed with dolomite) every week will help. For Toggenburgs and other dark goats, higher amounts of copper are needed, and finding the right amount to feed without giving her too much is a personal choice. Copper in dry treat feed is often eaten around by goats, so it is difficult to see just how much they are eating, it may be worth starting with a once per week teaspoon down the throat, as for Saanens, and observing her eye membranes, coat colour, and general condition, and increasing the dose slightly (or doing it more often than once per week), if there doesn't seem to be much change. Copper and other minerals can also be mixed into soaked grains after the water has been drained.

The tannins found in the leaves of some trees such as oak, maple, willow, birch, pine and yellow box, and also in chicory, grapevines, bird's-foot trefoil and big trefoil will help to prevent worms. If you don't have easy access to any of these plants, you can try feeding pine bark. Acorns will also help.

Regular checking of the eye membrane colour (see page 77) will help you to observe your goat's iron levels to determine whether she has a barber pole worm problem.

Coccidiosis

This is another kind of parasite; there are different forms that attack different animals, and chicken coccidiosis will not affect goats. Older goats can generally tolerate this parasite, but kids can have trouble with it. Having mallow growing (with plenty of other plants) where the kids are grazing, or by giving them a small amount to eat when you're feeding hay or other feeds will help them to become more resilient to this parasite. Mallow is said to be poisonous in large quantities, so make sure that when you are feeding them this that they have easy access to other food.

Preventing bloat

A goat's food is digested by bacteria in her rumen. These bacteria get upset by drastic changes in the food, and also by too much rich food, such as lush spring grass, or grain, along with a deficiency in potassium and magnesium. To prevent bloat (the buildup of gas in the rumen), make sure that before your goat gets access to any new feed in any big quantity, that she has filled herself up with her usual food, and that she has some cider vinegar in her water, and access to dolomite, either free-choice, or mixed into her 'treat' feed. Feeding her kelp also helps her to absorb the right minerals to avoid bloat.

One approach to avoiding bloat in practise usually works out to be giving your goat lots of hay to eat in the morning before grazing her on the lovely spring grass later in the day, or by introducing a new food gradually, so that if you suddenly have an abundance of grass or other leaves after a winter of hay, to at first give her lots of hay in the morning, followed by a short period of grazing, and then increasing the grazing time and reducing the amount of hay every day until she is not relying on the hay. It's important to observe her a lot if you are doing this, to make sure she can handle the new feed well. Sloppy poo is a sign that it has too much nitrogen, as is any sign of too much gas in her rumen.

It's important to observe her for any sign of discomfort and to treat it early. If you notice a slight buildup in gas in her rumen (usually on her left hand side) you can quickly supplement her by giving her a tablespoon of dolomite to swallow and adding extra apple cider vinegar to her water. Two teaspoons of olive oil down the throat will also help to relieve bloat in these early stages.

Transmissible diseases

The transmissible diseases discussed below become far less alarming when the goats are not deficient in any minerals. There have been cases of goats that get these diseases continuing to live long and happy lives once their mineral intake has been balanced.

CAE (Caprine Arthritis Encephalitis)

Caprine Arthritis Encephalitis (CAE) is passed from goat to goat through bodily fluids. CAE in unhealthy goats eventually leads to swelling of the joints, paralysis, and then death.

Sometimes there are no symptoms for a long time, so it can be hard to spot.

Vets can do blood tests on your goats to check whether they are free from this disease or not.

CAE is mostly transmitted from mother to kid through milk, but sometimes if a goat is injured and bleeding it can be transmitted through the blood. Whether CAE can be transmitted between buck and doe during servicing is still being questioned.

If one or more of your does has CAE, you can keep her kids CAE-free by taking them as soon as they're born (making sure they land on a clean towel without any birth fluids on it rather than the ground) and bottle-feeding them colostrum and milk from CAE-free goats, or pasteurised colostrum and milk from their CAE-infected mothers. The kids should then be raised separately from any goats that have CAE, and if you keep this up, eventually you will have no CAE in the herd.

Johne's

Johne's disease (also known as paratuberculosis) affects a goat's digestion. It is transmitted to kids through contact with manure from Johne's-infected goats, and it may possibly also be transmitted by milk. If a goat appears to have a healthy weight, it's unlikely that she has Johne's disease, but if she seems to have a good appetite, yet is losing weight and has diarrhea then she may have Johne's. A vet can do a blood test to confirm it. Johne's is very rare in animals that have been supplemented with the right minerals.

Tuberculosis and brucellosis

These diseases are around in some areas but not others. Find out locally from other livestock owners or vets if these diseases are present in animals in your area, if they are, make sure to get your goats tested for them, because they can be transmitted from goat to human by drinking the milk of affected animals. If in doubt, you can pasteurise the milk of any goats you suspect might have these diseases.

Brucellosis was found to be linked to deficiencies in cobalt, copper, iodine and manganese.

Preventing transmissible diseases

To begin with, you will need to get your existing herd tested for any disease you're concerned about. Once you receive a negative result, either keep a closed herd from then onwards, or only bring in animals that have already been tested.

The hardest thing for the backyard goat keeper concerned about livestock diseases is the need to get the does pregnant. This is when milking through can really help, especially if you have a doe that will happily keep lactating for several years, whether she gets in kid again or not. Keeping a buck is also a good option if you have the space for one; if not, then most show goat breeders that hire their bucks out will only service disease-free does, so you can generally be assured that their bucks are safe if you ask them about CAE and other diseases in their herd.

The most important thing any goat owner can do is to keep their goats healthy and in clean conditions by making sure they are getting the right minerals and food, keeping their food off the ground, and making sure they have clean bedding.

THE NEEDS OF
BACKYARD GOATS

Shelter

In theory, each goat needs 1 metre by 80 centimetres (3'3" by 2'6") of undercover space, around 1 metre (3'3") in height. In practise, some goats can be bullies and will not let other goats into shelters that have enough space for all of them in theory, and some goats might prefer to have a bit more distance from other goats, so it's best to either have a fairly large shelter, or have individual dog kennels for each goat.

A shelter must keep the rain, wind and snow off your goats, allowing plenty of ventilation without being draughty. It shouldn't overheat in summer. A three sided shelter, positioned so that the open side is away from most winds, is fine in most climates. If the shelter is not very tall it will need to be very strong, as your goats might want to climb and jump on it (kids are more likely to do this than adult goats). Shelter design will depend on whether you want to be able to move the shelter, or keep it in the one spot.

When your goats have kids with them you may wish to separate the kids from their mothers overnight, this is something to take into account when designing a shelter. This can be achieved cheaply and temporarily by having a small yard for the kids with a small shelter, and having the mothers in a separate yard with their own shelter. In a more permanent location a small barn can be designed with one area for the does, and another for the kids. There is a good design for a small shelter with a separate kid pen at goodlifepermaculture.com. au/urban-goat-keeping/

A hoop house or polytunnel design, with one end permanently open, and the sides reinforced with hog panels or another kind of rigid fencing is an affordable option if you don't have access to your own timber, where I live these are most commonly made from star pickets and 2 inch polypipe. Sepp Holzer's earth-sheltered log shelters, in his book *Sepp Holzer's Permaculture*, are a good option for building a permanent shelter on sloping land.

If your shelter is low enough that a goat can climb on top of it, keep it away from the fence, as your goats might use it to jump over the fence.

More considerations for shelter design are whether there will be an area for you to milk the goats incorporated into it or whether you will milk them in a separate building altogether. If you're designing a shelter to milk the goats in, it is best to have the milking area in a separate compartment, or with a gate to separate it from the goats so that they don't poo all over the milking stand. Another consideration that could change your shelter design is whether you will store the hay and treats in a separate compartment of the barn, or in a separate building. A place to provide mineral supplements for your goats to eat free-choice when they need to is a worthwhile consideration, as it's best to keep these out of the weather. I have never managed to do this, but if I ever build a barn, I will certainly make sure there is space for this.

If there is a lot of rain where you live, it might be a good idea to keep the hay feeders inside the shelter, out of the rain, so that your goats are able to get enough to eat without standing in the rain all day.

Fencing

If there's any way to test a fence, it's to put goats nearby. Goats are curious and intelligent creatures, if they want to get to the other side of a fence, they can often find a way. Goats also like to use fences and posts to scratch themselves, often running alongside it, pushing it with all their weight as far out as it will go, so a strong fence will gradually weaken from the goats doing this. Goats also like to climb up on the fence with their front hooves, which causes poorly made wire fences to sag, and possibly break. When a buck is in a rut he might butt against the fence in an attempt to escape and get to the does.

The less fence there is to damage, the more damage there will be, so a backyard goat paddock needs either very strong fencing, or electric fencing at the right height, with goats trained to stay away from it. A multifaceted strategy is best to deal with the escape antics of goats. The most important part of this is to keep your goats happy. If goats are well fed and their enclosure is carefully designed, they are less likely to imagine that the grass is greener on the other side. I've observed a doeling bouncing around our enclosure at great heights, and it was obvious that she could escape if she wanted to, but because she was happy she had no need to escape.

Raising goats in a backyard will usually involve a main enclosure, sometimes with temporary grazing spaces nearby where the goats will be either carefully tethered or fenced with portable electric fencing. For larger backyards of a quarter acre or more, you may be able to get away with rotational grazing all the time, with little or no brought-in feed.

The main enclosure should have very strong fencing, at least 1.2 metres (4 feet) high. Lots of straw on the ground will make the fence seem lower to the goats, so it's best to get a fence higher than this if you can afford to. Goats should always have plenty of feed, kept well off the ground (see what I've written later in this chapter about food for goats for more on this). The enclosure should have something to run around and climb on (a cable spool works well), and plenty of different places to run. Setting the shelter away from the edge of the fence gives them something else to run around.

Imagine the goat paddock as a place for children to play chasings or hide and seek and you'll come up with something suitable for goats. The enclosure is best if it's in an area that's comfortable to be in, ideally sheltered from strong winds, with some places to escape hot summer sun, and it shouldn't be in a frost pocket, or a flood-prone or boggy area. It's good if there's an easy and pleasant walk from the house to the enclosure, to simplify feeding them your vegetable scraps.

For a long-term strategy, it's worthwhile to plant trees that stand up to goats nibbling on them outside the enclosure, so that they will hang over the top of the fence, and through the fence to provide some very healthy feed. Trees which will withstand this nibbling include tagasaste, mulberries, elderberry, weeping willows, tree medic, some acacias, and leucaena. Another clever option for a goat paddock is to plant fenced-off hedges of tagasaste (tree lucerne) within the enclosure, or mature trees with sturdy guards around their trunks, so that the goats can't access the trunks, but can nibble at some of the leaves around the edges.

The size of the main enclosure will depend on how many goats you want to keep, how much space you have, and how much you can afford to spend on sturdy fencing. At a bare minimum I'd recommend 4 metres by 6 metres (13 by 20 feet) for two goats with kids up to three months old. If you're keeping the kids for a longer length of time, or think you might want a third goat in the future, I'd recommend at least 6 metres by 6 metres (20 feet by 20 feet). The more space you can give your goats, the better they will be for it and the more options you will have, so if you aren't moving house in a hurry it might make sense to just use a whole 50 metre (54 yard) roll of fencing and make the paddock as big as it can be.

Portable fencing

Another consideration for the enclosure is whether you think you will want to move it in the future. If you live in a rental property or think you may want to move house soon, an enclosure made from panels that can easily be assembled and disassembled may be the best option; these are often sold as dog enclosures. This option is usually more expensive than making an enclosure out of strong mesh fencing suited to goats, but it won't get damaged when you move it, as some other fences can.

A cheaper fence that is slightly more difficult to move (but still can be moved) can be made out of tall chicken netting or mesh goat fencing with metal star pickets (also known as Y-posts or T-posts). The star pickets are easily moved, and if you're lucky, the netting won't get damaged either. For this fence to work, the star pickets need to be spaced quite close together, every three metres at a minimum (closer than this is better). You will probably also need to use a thick wooden post for the gate to hang on, and to make sure that the mesh is well-secured to the star pickets at the bottom and at the top, to prevent the top from sagging and the bottom from being lifted up.

Another option for portable goat fencing is electric poultry netting, or in some cases 4 strands of "horse tape" style portable electric fencing tape. The netting usually comes in a 50 metre (54 yard) roll, and is more expensive compared to the electric tape, but much simpler to set up, with no tiny goat kids able to escape underneath. For 4 strand fencing, the portable insulated posts available are often low enough for goats to easily leap over, or in the all-in-one units the two top wires are often far apart enough that a smart goat will work out how to leap through, so you will need to either find taller portable insulated posts, or use metal pickets with plastic insulators.

Any kind of electric fencing other than the all-in-one kinds need special insulated stakes (or plastic insulators placed on metal posts); these are placed around every five to ten metres, depending on how flat the ground is, and the tape is wound through the hooks for it on the stakes. One very low strand of tape is needed to stop goats slipping underneath, and a strand of electric tape as high as it can go on the stakes is also needed, to stop them leaping over. Add two more strands, evenly spaced between these two, and you have a good electric fence for goats that are trained to electric fencing. You'll need to make a 'gate' in one part of the fence with plastic gate handle attachments. You can sometimes get away with using two of these with two strands of tape on each one if the 'gate' part of the fence is made just wide enough for a single person or goat to get through, but more is better. You will also need to link all the strands together by taking a small amount of tape and tying it onto each strand.

For any kind of electric fencing you'll need an earth stake, which is just a metal stake rammed as deep into the earth as it can be. You'll also need a good fence energiser connected to a reliable source of electricity. The energisers that I use are self-contained solar electric ones, which have a solar panel on top, with enough battery capacity inside to go for three weeks without any sunlight. Batteries are bound to fail at some point, and nothing electric will last forever, so the fence will need to regularly be tested, and possibly a backup battery or energiser kept on hand in case it needs to be replaced quickly. For poultry net fencing, a 2.5km energiser is recommended for 50 metres of netting, although it's always better to have an energiser that is more powerful than you need, than to find that the one you have is too weak to scare the goats away.

Test your fence every day to make sure it's going. You can do this with the back of your hand, or with a wet blade of grass. It won't always shock you right away because it runs on a pulse, so you may need to wait a few seconds to see if it works. If you're unsure about the strength of the shock, always test with the grass rather than your hand, as some people react badly to shocks directly on their skin.

Your goats may need to be trained to the electric fence. There are two ways to do this:

- If you have a small space to work with, tether your goat inside an electric fenced paddock, with enough slack in the rope so that she can reach the fence. Observe her and make sure she gets her nose to the fence and gets a shock; you can make this happen more quickly by bringing some food over to the fence. This is not something to do when you and your goat are getting to know each other - it's best to wait until you are established in her eyes as the bringer of food. Keep a close eye on her to make sure she doesn't get tangled in the fence.

- The second way is to make a smaller electric paddock inside a larger secure fenced area, it's best if the larger area has non-electric fencing, as she may not learn from the first shock. Poultry netting is better for this than horse tape, as it's more visible and difficult to get through.

Goats are intelligent and determined creatures, some goats will know that it's unpleasant to be shocked, but will risk this shock anyway in order to get at something they want to eat. I have found that pure Saanen goats will respect electric fences more than other breeds.

To help goats better respect the electric fencing, make sure that there aren't any sagging parts, or other gaps that might tempt them to leap over or sneak through. Some poultry netting is available with electric lines that alternate between positive and negative, so that the fence will give a stronger shock than a standard all-positive fence in dry weather. All electric fences will be less effective if grass or other vegetation is touching it.

Permanent fencing

To make a more permanent goat fence, you need to first decide whether you want to incorporate any electric fencing into it. Electric wires, usually placed slightly inward from the main fence, are placed at the top of the wires, and the bottom, to stop the goat from leaning on the fence with her hooves, and from trying to dig her way underneath it. Non-electric fences will need to be very resistant to a goat's weight. The available varieties of goat fencing, and the names they are called will depend on which country you live in, it's best to ask about it locally at your rural supply store, or to find out which brands of fencing are easy to order in where you live (or online), and contacting the company directly. Tall chicken wire is can be used if you are willing to use more stakes, but it won't last as long as the thicker-wired goat fencing.

Where I live, the kind of fencing mesh recommended for goats has 8 horizontal strands of wire for 90cm (35 inches) of fence height, with vertical strands of wire every 15cm (6 inches). The fencing is usually 90cm high, with two strands of electric or barbed wire on top to make it higher. When building a fence for goats it's better to overspend on stakes and have a very sturdy fence than to be stingy about it and find that your fence is sagging in places because the stakes aren't close enough together. The kind of stakes will depend on what's available locally. Metal stakes (sometimes called star pickets, T-posts, or Y-posts) can easily be hammered into the ground. Thick round wooden posts can also be used, but these will need a fairly deep hole dug for them.

Tethering

If you want to tether, it should always be done when you're at home and can listen out for the goat. They sometimes get tangled and can sometimes escape. To minimise the chances of these things happening, a sturdy collar needs to be on the goat which she can't snap or get out of, and a swivel clip needs to be used to attach the collar to the tether. If it's possible to attach a swivel to the object your goat is tethered to, do this as well. Even with these measures in place, a goat can sometimes still get caught in the rope, leading to discomfort and even injury.

Never leave a goat anywhere that dogs can get to her unless she has horns to fight back; even then it's a bad idea to have her tethered, as she can get tangled and be unable to fight back.

Contrary to misinformed popular opinion, goats don't make very good lawnmowers. They can be picky, and if there is a tree or shrub anywhere in their reach, they are far more likely to eat this than they are the grass; they can sometimes kill the trees by eating the bark off all around the trunk. If there aren't any trees or shrubs nearby sometimes goats will go for the grass and weeds, but they will often eat it unevenly, and sometimes won't be interested at all. Sometimes the grass is a good fresh supplement to their diet and helps with the lawnmowing, but in a small yard I've found the results vary too much to rely on tethering for feeding or lawnmowing.

Bedding

The grass will quickly be eaten and the ground compacted in the main enclosure. On flat ground, deep litter is a good solution. Deep litter involves adding more straw or other mulch on top of the goat poo; this can go on for 6 months or so before the whole lot should be removed and composted, and a fresh layer of straw added. On sloping ground deep litter won't work, so the goat pen should be raked out around once a week, with fresh straw added.

If you have sloping land, or don't want to use deep litter, try to use portable fencing or have multiple paddocks so your goats can be moved to a new area as often as possible, giving the earth a chance to rest and the parasite cycles a chance to die before they're returned to the same area. You could use these spaces while they're being rested for growing vegetables and crops, as a kind of 'goat tractor' system.

Food for goats:
Learning from wild goats

In the wild, goats are mostly browsers, rather than grazers. Wild goats eat mostly trees and other sources of food that are away from the ground; these plants often have deep roots which bring minerals up from the subsoil. We can learn from this and arrange it so that our goats can eat mineral-rich food which is off the ground in the backyard environment. When we learn from nature and apply it to the backyard environment our goats can enjoy healthy lives free from parasites and diseases.

Goats are highly prone to parasites if they eat their food too close to the ground. For grazing goats, the pasture should be at least 15 centimetres (6 inches) high if you want them to eat any of it. Hay and other brought-in foods will need to be kept off the ground, either with a hay feeder or manger, or with cheap hay bags or rubber tubs that can attach to the fence. Hay bags can be found wherever you can find horse supplies. Avoid the net style of them if your goats have horns, I recommend the style of hay bag in the photo on page 61; it's made from solid thick fabric, is filled with hay from the top, with a goat-head sized hole towards the bottom.

The strict butting order in goat herds, and the amount of time that goats spend eating means that you'll need to make sure each goat can access food when they want to. Providing a separate hay bag or tub for each adult goat will achieve this, as will a large manger with more than enough space for all your herd at once.

Goats can be very fussy when it comes to hay. Lucerne, clover, and other legume hays are usually easy to find, and favoured by goats (avoid red clover hay for white goats, it may be too high in copper for them). Second-cut grass hay, carefully made from fertile pastures can be acceptable. If horses are fond of the hay you're looking at, goats probably will be as well. Grain crops that have been baled as hay well before the seed heads are mature, while the grass is still soft, will also be favoured by goats.

If you're unsure about hay, it's best to just buy one bale and offer your goats some to see if they like it. Never feed them mouldy hay, and try to keep it out of direct sunlight, in a well-ventilated area. The traditional hayloft of a barn is the ideal place to store hay, it gets plenty of ventilation there. Garages and carports are fine for storing hay.

If you have the storage space and the money upfront it is worthwhile to arrange to get a year's supply at harvest time, rather than buying small amounts throughout the year, because some farmers will run out before the next lot is ready, and it can be stressful going on a wild goose chase trying to find another supplier. During our time living in rental properties with goats we always just bought two to three weeks supply at a time. We've had to put up with some very expensive hay at certain times of the year, and would have preferred to avoid this, but at least our goats got fed.

The amount of hay that your goats will eat depends on whether they have other food. If you're feeding them hay only, without grazing, two adult goats will generally go through between one and two small rectangular bales every week (usually around three bales every two weeks), depending on the quality of the hay and how tightly it has been baled.

The dry weight of food that a goat will actually eat is estimated to be between 3.5 and 5 percent of her body weight each day, so for a 65kg (143lb) goat, this works out to be between 2.27kg and 3.25kg (5lb and 7.1lb) of dry weight per day. All food has some moisture content, so the actual weight of the food will be higher than the dry weight. If your goat is eating a lot of scraps, fresh pasture, and leaves, it will work out to be more in weight than if she were eating lucerne only. If the food she's eating is low in nutrients, then she may eat more, and if the food is nutrient dense, she will eat less. Observation is always best – if your goat seems hungry, then she probably is. It's always best to allow free choice access to staple foods such as lucerne or tree branches, goats will adjust their own food intake depending on their energy and nutrient needs.

Goats are ruminants, which means they have four stomachs, one of which contains bacteria which ferment their food to digest it. Because of this bacteria it's important to allow goats time to adapt to any new feed given to them, and to feed garden scraps in moderation. Upsetting the goat's digestion by introducing too much new food at once can lead to serious health problems, even death.

Feeding scraps

Goats enjoy most fruit and vegetable scraps such as apple cores, orange peels, banana skins, the limp outer leaves of cabbage, broccoli stems, pumpkin skin and all the other bits and pieces of vegetables that we don't eat. They also enjoy scraps from bread and other baked goods. Goats should not be fed any meat, nor anything poisonous such as potatoes that have gone green, or anything mouldy. Individual goats seem to have different preferences for scraps; one of the goats I look after thinks that banana peels are the best thing ever but some of my other goats won't touch them.

Feeding trees

Goats greatly appreciate tree branches and leaves to eat. If you have access to suitable trees you may be able to get away with not buying in any hay at all, and instead bringing them branches. Goats generally love nitrogen fixing trees like acacias and tagasaste. They are fond of most maple leaves (although red maple is poisonous) tree ferns and some other species of ferns (not bracken though), willows, apple and pear trees. They like ash, elm, oak, poplars, and pines.

Goats have a good sense of what they can and can't eat, so if you're unsure about whether something is a suitable food for them, you can give them a small amount, with plenty of their usual food to eat so that they're not forced into eating the new stuff, and observe whether they are interested or not. I have found that goats prefer different trees at different times of the year, and that some goats prefer different plants than other goats.

In earlier times "Tree hay" was often made in the summer from ash, elm, ivy and oak. If you have storage space, you can cut small branches off these and other goat fodder trees and dry them in bundles hanging from the rafters. Nettles and other leafy plants can also be treated in this way. If the tree hay branches you collect are thin enough (around 1cm/1/2" thick) goats will often eat the whole thing, branch and all.

Goat fodder trees can often be found on public land, so if you don't have many trees at your own house, you can always go for a daily stroll with some secateurs or a pruning saw to gather branches.

Poisonous plants

Rhododendron and azalea are highly toxic to goats. Many other garden ornamentals are a bit suspicious as well, so make sure that before you feed anything to your goats, you've looked it up first to make sure it's not toxic to them.

There are lists online of plants that can be toxic in high enough doses, but goats will often eat small amounts of these with no problems. The key to avoiding poisoning is to always have plenty of food accessible that the goats will eat. Before feeding any new plant, make sure you've identified it and that it's safe for goats. Before tethering a goat, make sure that there's no rhododendron in reach, that bracken ferns have been thoroughly stomped down or removed, and that there is plenty of food within reach that the goats are eating at this time of the year.

Plum, peach, nectarine and cherry leaves can be toxic to goats, so try to make sure you don't have the goat paddock underneath the canopy of any of these trees. Some sources say that the autumn leaves of other plants aren't good for goats, but from my experience with sycamore maples, as long as there is plenty of other food around for them it's not a problem.

Growing food for goats

In a backyard it is best to either keep plants separate from goats and bring small amounts to them, or to offer controlled grazing, either by tethering them nearby for short periods of time, or by growing plants against the outside of their fence, so that some leaves can be accessed from the goat paddock but the goats can't gobble the entire plant up or eat all the bark off the trees.

Goats appreciate access to comfrey, either a couple of leaves offered in their feed each day, or being tethered near comfrey for a short time (with plenty of access to their other favourite plants in the same place).

Roses are good as a remedy for scouring, and goats also appreciate the taste of them.

A variety of kitchen herbs and 'weeds' can be grown and offered to the goats with other feed. Some herbs they might choose to eat at some times and not others, but a variety of food is good for their health, so it's worth growing a few extra herbs to share.

Soil pH and mineral availability

If you want to rely a lot on trees and other free foraged goat feeds, it can help to know the pH of the soil you are taking these from, and if it has any mineral shortfalls. If the soil is too acidic, or too alkaline, then the trees may be unable to access some minerals from the soil, and you will need to either give these to the goats separately as supplements, or remineralise the soil.

In general, a soil with a pH of between 6.5 and 7 will mean that plants can easily absorb all soil minerals well. With a pH of 6, plants can still absorb most minerals from the soil (only phosphorus is affected much), but between 6 and 5.5 potassium, sulfur, calcium, magnesium and molybdenum become less available. Soils with a pH of less than 5 will also have less copper, zinc, manganese and boron available. Alkaline soils, with a pH above 7.5 have less iron, manganese, boron, copper, and zinc available.

The subject of minerals in soil is a complicated subject, with excesses in some minerals blocking the absorption of others and so on, but it helps to be aware of how the acidity of the soil affects the plants that goats eat, and to supplement with the right minerals depending on your soil. For anyone feeding goats plants from acidic soils, dolomite lime for calcium and magnesium is probably the best supplement to start with, along with some cider vinegar added to the water for potassium. Sulfur can be fed too, which can be especially helpful if there are problems with the skin or fur.

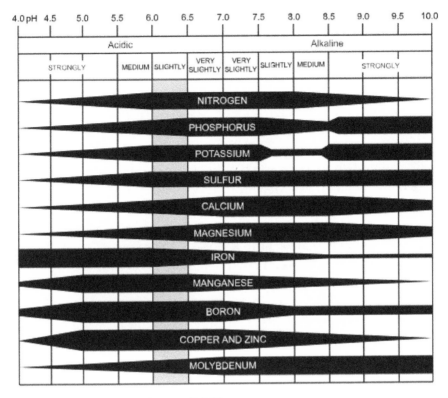

Effect of soil pH on nutrient availability.

Minerals and supplements

Copper is the most important mineral to add to goat diets in areas where copper is deficient in the soil, or where the water is high in sulfur, iron or calcium. Darker coloured goats have a higher need for copper than white goats, and you can often tell when a darker goat is deficient in this, as her coat will become lighter. Loss of hair on the tip of the tail, to give it a 'fish tail' appearance is another sign of copper deficiency. In 'Natural Goat Care', Pat Coleby states that she's never encountered a goat with worm problems when their diets have been supplemented with copper. Copper sulphate can be bought in animal feed stores, it's often found with the horse supplements. The easiest way to add it to the diet for milking animals is once a week to mix one teaspoon per goat into the 'treat' rations each goat receives while being milked, another way is to place it in small containers available to the goats to help themselves from, but you must make sure these are kept indoors or else they will be ruined every time it rains.

Copper oxide is more difficult to find than copper sulphate, but is a safer option for those worried about copper toxicity.

I had never heard anything about it being possible to feed too much copper until recently, and there still is no definite upper limit for it. I found when I was feeding my Toggenburgs copper sulphate every day (around half a teaspoon a day – over three times more than what I recommended above), they were very healthy and had no parasite problems. Here in Australia the soil is often low in lime minerals and copper, and our goats benefit from the extra copper, but if you have healthier soil, you may want to keep to a 1 teaspoon a week dose of copper sulphate, or to feed them copper oxide instead.

If sulfur is deficient in your soil or if you are noticing skin problems or external parasites on the goats, then it might be worth offering some gypsum or yellow dusting sulfur to the goats, either as a free-choice mineral, or sprinkled into their food.

Kelp (seaweed) is an excellent natural supplement that supplies a wide range of minerals, especially iodine, which is essential for the absorption of all other minerals and vitamins and especially important if you're feeding your goats lucerne. Kelp is best given free-choice, either by having a container on a wall out of the rain for the goats to eat as they choose to, or by offering it to the goats at milking time a couple of times a month to see if they are interested. I like to sprinkle a small amount on top of their treat-feed every day. Sometimes they will eat large amounts of it, other times they are either not interested at all, or will only eat a tiny amount. If they have never had kelp before, you may find they eat a lot at first.

Selenium is an important mineral for goats. Kelp, wheat, oats, and sunflower seeds are good sources of this (as long as the soil they're grown in is not deficient), so your goats may already get the right amount in their milking time treat. If you're concerned about the mineral levels of any goats that don't get grains, feel free to regularly give them a handful of sunflower seeds. Sulphur is needed to absorb the right amount of selenium, so supplementing with sulphur if the soil is deficient or acidic is a good idea.

Salt is essential for goats if you are not feeding kelp, but even then, you may find that they get enough of it from their regular feed. Goats know for themselves when they need salt and when they don't, so it's best to offer it free-choice (preferably as kelp).

Goats will sometimes eat a lot of salt when they are in need of potassium, so adding cider vinegar to their water can help add the extra potassium to the diet. Goats can be fussy about licking from a block of salt that another goat has been licking, so loose salt is preferred to blocks, or get one block for each goat. I have used Himalayan salt, but any natural unrefined salt without additives will do the trick. Instead of always having a salt lick there for the goats to lick at, you can just give them handfuls of course salt to nibble at every now and then, and they can either eat it out of your hand, or out of a bowl.

Calcium and magnesium are very important for dairy goats. Dolomite lime, either offered free choice, or around a tablespoon per day added to the treat feed will supply both of these minerals.

Potassium is important for pregnant does as kidding time approaches; apple cider vinegar added to the drinking water is a good source of this, and is a good supplement all through the year to boost immunity and digestion.

Pregnant and lactating goats need a supplement; usually this is a mixture of locally grown grains. Barley is said to be especially good for dairy animals, it increases the amount of milk. This supplement is best fed as a treat at the milking stand. Feeding a pregnant goat like this every day will get her used to coming to the milking stand and will make milking a lot easier once she has kidded. Avoid goat pellets at all costs. Goat pellets turn to something similar to mushy cardboard inside the goat's belly, and goats need more fibre than what pellets provide. Also try to avoid anything with molasses in it, or anything that is sweetened, as sweet foods make them more prone to insect attacks. If you want to buy prepackaged 'treat' feed, dairy meals designed for cows, made from cracked and rolled grains, can be an acceptable choice, or just buy a large bag of whole wheat and a smaller bag of sunflower seeds and mix them together, or even just plain wheat or plain barley is good. Soaking wholegrains in water with a splash of cider vinegar overnight or for 24 hours will enhance their nutrient availability and make them more digestible. I usually soak one batch of barley in the morning each day, some of it is fed in the evening, and the rest of it the next morning. Before feeding the barley, I drain the soaking water and then mix in the daily rations of copper sulphate, yellow sulphur and dolomite lime to the grain.

Before you get your goats, research soil mineral deficiencies in the area you'll be buying hay from and try to offer these minerals as free-choice, or carefully add small amounts to their treat feed. Offering minerals free choice makes it easier for the goat to correct her own nutrition when she needs to, but you'll need a way of keeping these out of the rain. If the goat isn't interested in the minerals to begin with, you can try sprinkling a small amount of grain over the top of them to encourage her.

For Australia, where our soils are mostly acidic and deficient in copper, Pat Coleby recommends a basic stock lick made from 6kg (12lb) of dolomite, 1kg (2lb) yellow dusting sulphur, 1kg (2lb) copper sulphate, and 1kg (2lb) kelp. These minerals can be found in animal feed shops and horse supply shops. Dolomite is easily found in any garden centre.

Hoof trimming

Goats are from mountainous areas where their hooves are worn down from daily wanderings and jumps on rocky ground. We can imitate this to some degree by having some large rocks for the goats to climb on in their paddock, but you still need to keep an eye on their hooves, which generally need to be trimmed every eight weeks. Speciality hoof trimming shears (also called footrot shears) designed for sheep or goats are the best tool for this job, but garden secateurs can also be used, or a sharp knife if you have enough confidence and a goat that stays still. If you don't trim their hooves on time, the hooves can grow very long and curl around over the base of the foot, trapping mud and goat poo, which may rot and cause health problems.

When trimming a hoof it's better to start by trimming a small, even slice, all around the hoof first, then trimming another small amount until you are very close to the foot, and everything appears to be even, clean and comfortable. It's possible to trim too closely, and the goat can get cuts on her skin from doing this, so it's better to try a little at a time, and if there's any doubt about whether you've trimmed enough off or not, it's better to err on the side of caution and not trim off any more. It's good to do this on the milking stand with a bowl of treats for the goat to eat. Depending on the personality of your goat, you may need someone to help hold her leg still. You may need to do the hoof trimming over two or four days, especially if you've been milking the goat on the stand before you start, as she might decide she's eaten enough treats and that it's time to go back to the paddock.

The milking stand

This isn't essential right away, but it will make life a lot easier when you need to milk your goats or trim their hooves. There are free instructions available online for making them out of pallets and other wood, and they can occasionally be picked up second hand from sellers on Craigslist, Gumtree and other classifieds. A good milking stand will have a means of securing the goat to the stand, usually by having her head go through an opening which can be closed into a size big enough to be comfortable around her neck, but small enough that she can't move her head back through. Another way is to have the goat on a leash, and to secure the leash to the milking stand or to a wall next to it. For a photo of my milking stand, see page 63.

Water

Goats need clean water available at all times, and for a small backyard herd this is easy to manage. I've found it easiest to provide this in sixteen litre (three gallon) buckets, I use one or two in winter, three or four on hot summer days. It's important to check up on it twice a day to make sure they have some left, and that it hasn't frozen in winter. One of these buckets we attach to a clip at the end of a rope attached to the fence. My husband can reach over the fence to lift it up and down using the rope, so that we can easily refill the water without going in and out of the paddock. Refilling the buckets via a watering can is another quick option, but you'll still need to remove the buckets when they need to be cleaned. Rubber or flexible plastic tubs designed for horses work well for goats. Cheap plastic buckets can be used, but will not last long.

Always check to make sure your goats haven't pooed in the water; if they have, it should be changed right away. It's best to keep buckets in a couple of different places in the paddock, so that it's less likely for them to both be knocked over or pooed in at the same time.

If you live in a climate with very cold winters, you will need to either insulate the water container or use some sort of system with a heater.

Goats appreciate warm water in winter and cool water in summer; on very hot days you can put ice cubes in the water, and on cold days they will appreciate a bucket of warm water if you can manage it.

A goatherd's responsibilities

Before you get goats, you'll need to consider whether you are available to be responsible for your goats. Responsibilities of a backyard goatherd are:

Being available to milk her, either once or twice a day
For some goats during early lactation, if she has no kid with her, you

may need to milk her twice a day, but most goats generally cope well with once a day milking.

This doesn't have to be at the exact same time every day, or exactly twelve hours apart if it's twice. Once your goat has settled down and is being milked once a day (if that's what you've chosen to do) you can go away on holiday for a night, milking her early before you leave, and milking her late when you return. She will prefer it if you milk her at around the same time every day, but exact accuracy is not essential.

If you can't be there every day for your goats, you'll need to find someone who can look after your goats for you while you're away; or you could consider looking for someone who wants to co-own the goats, milking some days while you milk on other days.

Responsibility for the kids

Before any goat has her kids, you need to work out what you will do with them. If you're planning to sell them, then you still need a plan B in case they don't sell, either eating them, or giving them away. You will need to decide whether to turn the buck kids into wethers, or to keep them intact, keeping in mind that it's usually harder to sell an intact buck kid than a wether, and if your plan is to eat them, bucks develop a taste that not many people appreciate. It's much easier to wether a young kid than it is a larger one, so it's something that will need to be thought about early. For more on wethering see page 90.

You will need a plan for how to deal with the kids drinking their mothers milk – will you keep them with their mothers full time until they're sold, accepting that you won't have much milk for a while, or will you keep them with their mothers during the day and separate them at night so that you can milk the does in the morning? If you're separating them at night, then you'll need a secure area for the kids where they can't get to the does. Some goat owners take the kids and bottle feed them from the beginning; this gives you more control over the milk, but is a lot of responsibility, and the kids will not be as healthy as those raised by their mothers.

There is also the chance that a mother will reject her kid. If she gives birth undisturbed and is comfortable in her surroundings, then her instincts will guide her to look after the kid, and it's unlikely that she will reject it unless there is something seriously wrong with it, but if you're overly eager to 'help' the doe during the birth (or if she genuinely needed help), then you may find yourself stuck with a kid to bottle feed, whether you wanted it or not. For more information about how to correctly behave at a goat birth, see page 84.

Your goats need you to protect them

If you are near a road, you'll need a fence that will keep the goats away from it. If there are savage dogs or other larger predators around, then you'll need a fence that will keep them out, and to consider getting goats with horns, so they can better defend themselves. If you're separating the kids at all, you'll need to consider any predators that may attack them. You might consider getting an alpaca, llama, shepherding dog, or other livestock guardian animal if there are a lot of goat predators where you live.

You will also need to keep the goats out of anywhere they're not meant to go

Good fences make good neighbours. Your yard might be close to your neighbours house, but they may not like goats as much as you do, and if they find escaped goats roaming around eating their prize rose bushes they might complain to someone who can send you a letter full of legal jargon to tell you that goats are not allowed in backyards because the area is zoned as such and such. Neighbours will put up with the goats more readily if your goats are securely fenced in and never escape. For more on fencing, see page 31; for more on dealing with neighbours and landlords see page 52.

Goat health

You are responsible for observing your goats and making sure they are in good health. You'll need to trim their hooves every couple of months, provide them with enough feed in a way that avoids ground-dwelling parasites (see page 38). You will need to provide any minerals that are deficient in their feed. If your goats end up with health problems, you will be responsible for helping them.

GETTING
YOUR
GOATS

Before you get goats, it's important have read the earlier parts of this book, to understand what you will need to have set up in advance, to organise food supplies, and to be confident that you can give them the care that they deserve.

Dealing with neighbours and landlords

Many people, when asked to give permission for something that seems a bit unusual to them will often give an automatic 'no', or they might imagine goats gobbling up their rosebushes and then say 'no'.

If you own your home, all you need to do is to research your rights and see if a neighbour has any legal standing if they tell you to get rid of your goats based on council regulations. If there are no council restrictions to goats, then you can go ahead and get them; if there are, you might want to be on good terms with your neighbours first, just to make sure that no one is going to dob you in for having goats.

Every situation will be different, the most difficult one to manage is if the only place you can put your goats is close to where a neighbour's house is - this is the kind of situation where it is appropriate to discuss it with them, but if you have a large yard and the goats are going to be well away from anyone's house, then you can probably just get away with setting things up and if you find someone peering over the fence looking for something to complain about, you can kindly remind them that your goats don't have any effect on their life.

Typical concerns from neighbours:

"What if they smell?"
Bucks can be a bit stinky, but female goats don't stink. Any smell from manure happens only when nitrogen is escaping to the air; if you're able to keep plenty of fluffy straw around to absorb the manure, or to remove the manure each day to put in the garden or compost heap, then there won't be a smell.

"Aren't they noisy?"

Goats bleat every now and then, but they won't wake you up or distract anyone unless it's someone who really hates goats. European goats have a quieter bleat than Anglo-Nubians.

"It's unhygienic."

Goats are no less hygienic than a dog or cat. Goat poo is dry and doesn't smell, and can be easily managed to be out of sight.

Renting

With landlords it is a different story. There might be a clause in your lease that says that you're not allowed to keep animals, and you can be evicted if you break that rule, so if this is the case, then it helps to discuss it with the landlord, and to let them know that you would only be keeping the goats on the grass (or on some other part of the yard they aren't too fussed about), and will remove all traces of the goats having been there when it's time to leave.

To avoid the stress of suddenly having to move house with goats, it's worth finding out first what the landlord's intention for the house is, if they are planning on selling it soon then you might be faced with having to quickly find a new place. The hardest thing I found about keeping goats in a rental property was when we needed to move house at a time when there was a shortage in rental houses. The landlords and real estate agents had a task of finding reasons to rule someone out for renting a house when several applications were in, and goats were one of their things to nitpick about. If you're moving house during a time when there are lots of houses for rent and not much demand, you might find that you don't have this kind of trouble, but in general, keeping goats can mean that it will take longer to find a rental house.

The two situations I rented houses with goats in were similar – old houses with mostly lawns around them in the middle of working farms. The farmers would rent out these cottages, which were nowhere near other neighbours. This is a fairly common situation now; farms in the past used a lot more human labour than they do today, and cottages were needed for the workers, but these days they are rented out for the farmer's income, with less danger of being suddenly sold than a standard 'investment property' that changes owners often.

Country towns and rural areas are usually more tolerant of goats than the suburbs of cities, because it is 'normal' to keep livestock in these areas. As more people everywhere become aware of sustainability issues such as plastic waste, conventional dairy practises, and the effects of the current agriculture system on the environment, they will become more accepting of home dairy. It was not that long ago that keeping chickens was considered 'abnormal', but now that is seen as acceptable, the typical suburban perception about dairy goats should hopefully change for the better as well.

How much milk?
How many goats?

The amount of milk you can expect from one dairy goat depends on her genetics and how she's been raised. Goats who have become pregnant too early (see page 78) often have udders that haven't developed properly, and they will give less milk. An overweight goat will also give less milk. A good udder should have a rounded shape with plenty of space for milk. Droopy and badly attached udders can still produce good amounts of milk, but are more prone to injury.

If your goat is from a line of full-sized dairy goats that have been kept for milk, has a good udder, and is given unrestricted access to lucerne or good hay, then she will probably give between one and two and a half litres (one and two and a half quarts) every time she is milked. Some goats will give a lot more than this. If she is from a line of dwarf goats that have been kept for milk, she will give around one litre (one quart) per milking, sometimes half this amount, sometimes up to two litres (two quarts).

Most goats will give a bit less milk over winter, and more in summer. If you want lots of milk and don't mind milking twice a day, then you can easily double these amounts to get the total amount for each day.

When deciding how many milking goats to get, considerations are:

- Minimum amount of milk that you need.
- Amount you would ideally like to drink, if you had an abundant supply.
- Whether you would like to make cheese, yoghurt or any other dairy products, and how much of these you would like to eat.

If you want to decide the amount of goats to get based on the minimum amount of milk that you need, and the goat is a good dairy breed, base it on the lower amount of milk quoted above (one litre (quart) per milking for standard breeds, half of this for dwarf breeds.) Any extra milk you get will be a bonus.

If you would like to have cheese as a staple food, to eat meals based on ricotta and paneer, or to enjoy plenty of hard and soft cheeses throughout the year, a good rule of thumb is that for every kilogram (2 pounds) of hard cheese you wish to make, you will need eight to ten litres (8-10 quarts) of goats milk, for every kilogram (2 pounds) of soft cheese, ricotta and paneer, you will need around six litres (6 quarts). Milk shouldn't be kept for too long before being used in cheeses, I generally make raw cheeses within 3 days of milking. Ricotta and paneer can be made later than this, as long as the milk still smells good and is kept very cold so that it doesn't begin to ferment on its own.

To calculate how many goats you'll need based on cheeses (and my fresh milk to cheesemaking times), you'll generally need 3 low producing goats in milk to make a hard cheese within 3 days of milking if you're milking once per day. If you're milking twice per day this number can be halved. For the average goat, if she's from a good dairy line, you'll need two goats to make this if you're milking once per day, one goat if it's twice.

These numbers don't take into account the amount that you'll need for drinking. There is also the option of halving cheese recipes and using smaller cheese baskets, so that you can get away with using four or five litres at a time for hard cheese rather than eight. I've had good results making cheese in six litre batches in an 800g cheese mould (see my Tomme recipe on page 128). Soft cheeses are best made in smaller batches, so it's possible to make some with just one or two litres.

Horns or no horns?

If your herd has a mix of horned goats and goats without horns there may be injuries due to the way that goats naturally fight. There are some exceptions to this rule, but in general, unless you know how a certain goat has handled being around other goats, it's best to keep horned goats separate from dehorned ones.

It's important to decide from the start whether you would like goats with horns or without them. Goats naturally have horns in the wild; the horns help to regulate their internal temperature and improve general health.

Tests done on cows with and without horns found that the milk was of a higher quality from the cows with horns. It's helpful to be able to grab onto horns if you need to catch a runaway goat, and as something to hold onto to lead them sometimes. From my personal experience selling goat kids, I found mine were sought after because they had horns.

Some people are afraid of goats with horns, and some goat books advise against getting goats with horns, but I find the benefits of horns outweigh the risks, and all that's needed to negate the risks is to have a good relationship with your goats, and to be aware and alert when you're in the same paddock with them. If you are afraid of horns, then you'll probably have less stress with your goats if you get goats without them.

Sometimes it's difficult to find goats for sale that have horns, so it helps to be flexible about it. One of my goats has no horns and we found that after a settling in period where she was on the other side of the fence from the herd for a long time, they eventually adjusted to each other and there aren't any injuries now.

Dairy goat breeds

If you're keeping a goat for milk, it's best to get a breed known for producing dairy. These goats tend to be skinnier than meat breeds, as they put their energy into producing milk instead of putting on weight. A good dairy goat will have a wedge shape to her body when viewed from above - her shoulders will be narrow and her body will gradually widen to begin with as if there is an invisible wedge running flush with most of her body, from the shoulders to the rumen (see photo on page 62) for an example of this).

Meat goats tend to have bulkier shoulders. Any female goat will produce milk after she has had a kid, but a dairy goat will produce more of it, and will generally lactate for a longer time than a meat goat, making it easier to continue milking her for years without needing to get her in kid every year. The dairy goat breeds you can find will vary depending on which area you live in. The most common in most areas are:

Toggenburg. A Swiss breed with a brown body and white 'stockings' on their lower legs and white markings on their faces. They have the archetypal European goat shape to their faces and horns. Toggenburgs generally produce good quantities for milk, over a fairly long period of time, making them a good choice if you don't want to get them in kid every year. Geraldine on page 61 is a Toggenburg.

Saanen. These are similar to Toggenburg, but they are usually larger (the largest of the dairy goats), and they are all white in colour. Sunshine, on page 62 is a Saanen. Some rare colour variations exist from recessive genes; these Saanens are known as sables, and except for their colour are the same as other Saanens. Melaans are black Saanens. Saanens produce a good amount of milk, milder in flavour to other breeds, but can sometimes reduce their milk supply unless they're kidded every year. Saanens are generally the easiest goats to manage; they have a reputation for being a bit stupid, but this means they are easier to contain in a fence and lead around, so is not always such a bad thing! They can have very sweet personalities and are a good choice if you are after a goat that is predictable and easy to handle. Saanens tend to have twins and triplets more than other breeds, so it's often easy to find good ones because many Saanen owners end up with too many goats for their own needs.

If you have a very shady yard, a Saanen is a better choice than a darker goat, as she will be more efficient at converting the available sunlight into the vitamin D that she needs and be healthier as a result.

British Alpine. These are similar to Toggenburg in appearance, but are black instead of brown. They often have more of a stockier build about them than a Toggenburg, and are known to be the hardiest goat in cold climates. Some British Alpines are known for producing milk year after year without needing to be in kid again. Geraldine from 'The Good Life' English comedy series is a British Alpine. The goat kids on the cover of this book are ¾ Toggenburg ¼ British Alpine.

Alpine, sometimes known as French Alpine. These are a reliable European dairy goat that come in many different colour variations, usually one colour towards the front, shading to a different colour at the back, with black 'stockings' on their lower legs. The Oberhalsi breed from Switzerland is similar to Alpine.

Anglo-Nubian. These goats originated in Africa and are better for hot climates than the European breeds. Made famous as a mystical hermit's source of milk in the dystopian novel 'The Heretic Emperor', these goats produce a good quantity of milk high in butterfat, making them a worthwhile choice for anyone who wants to separate the cream from the milk to make butter or use as cream. Anglo-Nubians can be a lot louder than the European breeds, so are not a good option if your backyard is close to other houses, or if you need to be discreet. They are mostly black in colour, with some brown and white patches, they have floppy ears and a convex nose. Anglo-Nubians are seen by some people as a "dual purpose" breed, producing more meat than the other dairy goat breeds, while still giving a good amount of milk.

La Mancha. These goats are known for their tiny ears and are readily available in a lot of places in the USA, but not in other countries.

Nigerian Dwarf, Australian Miniature Goats, and other dwarf goats. These are popular goats to be kept as pets, and it's often hard to find one that has been milked before, so it's hard to ask the owners specific dairy questions (although some of them will know, so it is worth asking). The standard of the breed can vary a lot, as many breeders are not looking for the qualities that dairy goat breeders look for. They tend to be more expensive than the other dairy goat breeds. Advantages they offer over other goats are that their kids are often easy to sell as pets, and that they need less space than a regular goat (three dwarf goats can be kept in the same amount of space as one full size goat), so if you have a very small space or very limited feed budget you can keep more of these goats on your land than the larger breeds, and their small size makes them easier to handle. Dwarf goats usually give less milk than other dairy breeds, and the small teats on most of them can make them more difficult to milk.

Other dairy breeds
There are many other breeds of goats around the world, some of them suitable for dairying. You may find that there is a breed of goat well adapted to your climate and local conditions that can easily be found where you live.

A note on crossbreeds
If a dairy goat breed has been crossed with other dairy goat breeds, then there won't be much of a problem as far as milk production goes. But when a dairy goat has been crossed with meat goats or generic 'pet' or 'lawnmower' goats, then her milk supply is likely to be a lot less than a pure dairy goat's, and she may not be capable of the long lactation that home dairy goat owners look for. There are some exceptions to this rule, but unless you know the history of a particular goat, then it's worth waiting until you can find a good dairy goat.

A breed for each purpose:
For meat and dairy: Anglo-Nubian
If you prefer to sell the kids as pets: Miniature goats
Easy to handle: Saanen
Long lactations: British Alpine, Toggenburg
For thicker yoghurt or making butter: Anglo-Nubian
For hot climates: Anglo-Nubian
Quieter: Any European breed
For cold climates: Any European breed
For the tropics: Any dark-skinned goat, preferably Anglo-Nubian
For shady yards: Saanen
For lots of kids: Saanen

Signs of a quality goat

Have a look at the photo of Geraldine on the opposite page. Her face looks alert and healthy, her back is straight, her legs look as though they are supporting her well. Her udder is well-attached and a good size. I think that she is the perfect home dairy goat.

The goat on the right side of the picture on page 62 is Sunshine, a pure Saanen descended from a herd that used to be registered. Her mother gave more milk than any other goat in her herd.

If you look at Sunshine from this angle she has the perfect dairy goat 'wedge' shape to her, where if you imagine an invisible line going from each shoulder to her rumen, and continuing on at the same angle, you will see a wedge shape. The large barrel-like shape to her rumen obviously has a lot of capacity for digestion and breathing, and later on in her pregnancy this will mean she has more space for babies, and her udder will most likely be a good size.

The photo of Geraldine on the milking stand on page 63 demonstrates the best leg angle for a dairy goat. There should be a curved shape from the goat's tail to her knee.

Geraldine in the foreground. Her udder is well-attached to her body, has a rounded shape, and reaching to her knees has plenty of space for milk, without being so big that it can be easily injured.

Dairy 'wedge' shape of a pure Saanen doe. Sunshine is from a line of pure ex-registered Saanens. Sunshine's mother gave more milk than any other goat in her herd. Large barrel-like rumens like the one above can be an indicator of good milking qualities.

Geraldine on the milking stand.

How to buy your goats

Depending on where you live, your budget, what breed you want, and how far you're willing to travel, you may have quite a few goats to choose from, or you may have to accept the first dairy goats you can find.

Before buying a goat it's worth keeping in mind how long you will need to look after her before she starts giving milk. It's usually best for a goat to have her first kid when she is around 24 months old, so buying a goat much younger than this will mean a long wait. Buying a goat that is already pregnant or in milk is usually the best option, as it saves the hassle of finding a buck.

If you're buying a goat that is in milk and doesn't have any kids with her, be very careful that she has a docile temperament and will let you milk her, so that she can easily be milked on the first day she arrives, and you don't find that she dries up while you're getting used to each other.

Goat breeders (or goat studs) charge a fairly high price for goats that are 'show quality' and registered as purebred, but may also have some that are not quite up to standard that they will sell for less. Unless you are planning to enter your goats in competitions, you don't need a show quality goat. These goat breeders have often screened their goats for transmissible diseases such as CAE and Johnes, and they look after them much better than someone attempting to use them as lawnmowers, so if you can afford to buy from a breeder (and if they have does for sale), it may be worth getting quality goats to begin with. Another advantage of buying from a breeder is that they probably have a buck you can use in the breeding season.

Goat dairies will sometimes have goats for sale. Each dairy will be different, but might regularly sell their does once they get above a certain age, or they may also have some for sale that aren't quite up to commercial standards but will still produce a good amount of milk for a home. Sometimes milking machines can damage a goat's teats, so if you can, ask to hand milk the goat, to make sure that she still can be hand milked. It may require some adjusting where you would normally put your hands, and the milk may not flow out as well as a non-commercial goat's milk would flow.

Online classifieds websites and local papers can yield a variety of goats, from the great quality descendants of registered purebred goats, to the offspring of some random mixed breed goats that someone has in their backyard and has never tried to milk. It helps to ask questions to get an idea about how much milk you'll get, but there are good goats out there whose owners have no clue about milking ability.

Some things that will help you decide whether a doe will give enough milk are:

- When did she first give birth? Did the owner get her in kid in her first year, or wait until later?

  Some goat owners see no problem in getting does in kid as soon as they are physically able to, but this can stunt the doe's growth and result in a smaller udder capacity than she would have if she had waited. If you are buying her during this first pregnancy you will find that she is more susceptible to parasites, and may have a difficult birth and recovery.

- Where were the doe's mother and father from? And does the owner know about the milking history of the mother?

  If the doe's parents are from a dairy goat breeder, she will probably be a good home milker.

- If she is currently in milk, you can ask her owner if the kid can be separated for a few hours before you look at her, and that way you can milk her to see how much she gives, and how easy she is to milk.
- It is also worth asking how easily she kids, or how easily her mother kids.
- Has she been handled much?

If you're not sure about whether keeping dairy goats is a long-term choice for you, or if you don't need much milk, it may be worth starting out with whatever cheap dairy goat you can find, and then finding better goats later on if you feel more committed to it then.

Sometimes there are hidden gems among unknown goats for sale, some people do keep dairy goats as pets and have no idea about milking ability, and a really good one can sometimes be found in these circumstances. Generally it helps to familiarise yourself with what a good goat looks like, so you can look at them and see by the width of her ribcage and her shape if she's likely to be a good milking goat or not.

It's important to find out if your goat has been handled much, and if she'll follow a bucket of food, and if she is happy to be led around by her collar or a horn. For your first goat it's worth getting one that is easy to handle - someone's pet goat that has been patted a lot and showered with treats, or a docile commercial dairy goat are usually good choices for this. Wilder goats can be tamed, but it's best to have some experience with other goats before attempting this. If you are able to, spend some time with your goats before buying them, and observe how their owner gets them to move.

How to bring your goats home

It is easy to transport one goat at a time in the back of a car. The boot section of Landcruisers seems to have been made for this purpose. For smaller cars, if you don't have anyone sitting in the back seats, a goat can usually be convinced to go through the side door and sit or stand there, especially if you lure her in with food.

A fully grown goat can be heavy, so it's good to have a second person there to help lift the goat up if your car is high off the ground or if the goat is reluctant to go in there herself.

Always secure your goat by securely tethering her to a part of the car, such as a part of a seat, or a handrail, preferably with just enough room so that she can choose to sit or stand, but not enough room to go roaming through the car towards the steering wheel or get tangled.

Horse floats and trailers with high sides can sometimes be hired, which makes it easy to bring home more than one goat at a time. It's good to tether your goat inside the trailer, so that she doesn't try to jump out in the traffic, and so she doesn't run out as soon as you open it at home.

If you have a large car, you can sometimes put a row of seats down to fit more goats in. The challenge of transporting more than one goat is handling them when they are being loaded in, and when they get home.

Individual tethers, done in a way that won't distress the goats or tangle them, are a good idea if you can find enough places to attach them.

If you are travelling a long distance, try to do it on a day that isn't too hot. It's difficult to get water to the goats when they're being moved, so they can sometimes get dehydrated on a long drive on a hot day. When you get your goat home, offer her plenty of water and feed, in a non-stressful environment. Don't introduce her to the herd right away, or there may be too much fighting for her to have a chance to drink or eat.

DAY TO DAY GOATKEEPING

Goat health symptom checker

To get the most out of this symptom checker, observe your goats often to understand what they look like and how they behave in different weather when they're healthy. If you do this, then you will pick up on any issues quickly, when it is easiest to fix them.

Bulge on her left hand side
This is most likely bloat. It can be serious, or mild. She will always benefit from the magnesium in dolomite lime if she has a serious case or a mild case. She will also benefit from being given access to plenty of non-rich feed, although she may be off her feed and not quite herself until the bloat has been resolved.

Bloat is when the goat's rumen becomes filled with gas that can't escape. It happens due to a big change in feed, to too much nitrogen-rich feed (such as in the springtime when the grass is growing beautifully after a winter of dry hay) or too much grain. Her left side will bulge if she has bloat, and she will appear to be uncomfortable. Feeding her a small handful of dolomite by opening her mouth, shoving it in there, and then holding her mouth closed long enough for the inside of her mouth to be coated in the dolomite will give her the magnesium that she needs to balance her minerals and get rid of the bloat. Keep a close eye on her, and repeat this dose two or three more times throughout the day if you see no improvement. Apple cider vinegar will also help.

Sloppy poo
Healthy goat poo should come out as pellets. When it appears in one piece, or as runny diarrhea, then there is a problem, either a mild one which will resolve on its own (but will still benefit from the solutions in the next paragraph), or a serious one. If it's mild, then she will be behaving as she usually does, if it's serious, then she will be noticeably miserable and possibly not interested in food.

It could be internal parasites, or it could be that she has had too much rich green feed at once, or a big change in her feed. A small handful of diatomaceous earth in her mouth is worth trying, repeating two or three times a day if she doesn't improve. A chopped up clove of garlic or two given at the same time is also a good idea. If you can access tannin-rich leaves (see page 25), feeding these will also help her deal with internal parasites. Giving her the right minerals will also help. A lack of copper will make goats more parasite prone, as will a lack of vitamin B12 which is caused by a cobalt deficiency in the soil. Feeding her vitamin B12 powder or tablets will help, as will vitamin C. Observe her carefully, she may be off her feed and drink. You can encourage her to drink by gently holding up the bucket to her mouth, so that she gets the water on the end of her mouth and can lick it off; hopefully she will then begin to drink by herself. Never force her to drink or the water may end up in her lungs. Keep feeding her the same minerals and diatomaceous earth two to three times a day until she improves. If she is off her feed for 24 hours, help her by putting small handfuls of easily-digested food in her mouth. Avoid grain until she is better.

If a kid has sloppy poo (scouring), they can become dehydrated more quickly than adults, so it's even more important to observe and to encourage them to drink water. Hold the bucket up to the mouth to encourage drinking, and do this several times a day if the kid is not drinking on its own, but do not pour any water down the mouth. Half-doses of the vitamins mentioned above will help, along with diatomaceous earth and garlic. If the kid isn't drinking water yet, encourage him to drink milk by holding his mother still while his head is close to the udder.

Green or yellow mucus on the nose
This needs to be taken very seriously as it could be pneumonia. If it is pneumonia, it may be accompanied by raspy breathing, and your goat not being her usual self. Vitamin C tablets crushed up and fed to her by holding her mouth open, putting the tablets in, and then holding her mouth closed so that her saliva begins to dissolve them before she has a chance to spit them out will help. Garlic and ginger, chopped into small pieces and fed in the same way will also help, as will dolomite. Pneumonia is more likely to occur when your goat is under stress from pregnancy, a recent birth and/or nutrient deficiency. It's more likely to occur after the weather quickly turns

from a patch of cold, wet days into warm days, but it won't happen unless the goat has underlying nutritional problems or other stresses, so make sure from now on that she is regularly getting kelp, and any minerals that are deficient in your soil.

Try to keep your goat in a warm but humid environment while she recovers, and give her warm water with apple cider vinegar to drink. A few leaves of horehound added to the water will also help.

Raspy breathing

Make sure her collar isn't too tight, if not, then it could be pneumonia, see "Green or yellow mucus on the nose" on page 71 for treatment.

Grinding her teeth

This will most likely be happening because she is in pain. She might be in labour, or she may have another problem such as a bellyache. Observe to see if there are symptoms of anything else such as a bulge on her left hand side, or sloppy poo, and add some cider vinegar to her water. Don't give her grain until the issue is resolved.

Lowered head, leaning head against objects

This is also a sign that she's in pain, see "grinding her teeth", above. If she just has a lowered head and isn't leaning it on anything it may be that she doesn't like the weather, or has a bit of a goat cold. Feeding vitamin C and dolomite, and adding cider vinegar to the water will help.

Lying down, unable to stand

It's important to know the difference between a goat being in a bit of a lazy mood, or a bit under the weather, and being unable to stand. You may need to gently pull her to her feet to find out. If she is just being lazy or a bit sick she will stand for a while; if she has very bad bloat, she won't stand at all. While you pull her up, observe to see if the left hand side of her rumen is bloated. If she is bloated but doesn't seem to be in much distress, you can give her a tablespoon of dolomite down the throat, 2 teaspoons of olive oil also down the throat and some cider vinegar diluted by water, held up to her mouth for her to drink by herself. If you are very worried you may need to either call a vet, or use a sharp heat-sterilised knife* to make a small incision in the right place to get the gas out of her rumen. The correct place to do this is four fingers from the end of the ribs on her left hand side.

(*To heat-sterilise a knife, submerge it in boiling water for at least three minutes, or heat it in an oven with a temperature of at least 100°C (212°F) for longer.)

Pale around the eyes

If you are regularly observing the membranes under your goat's eyes (see page 77), you will notice if there is any change. White or pale pink membranes around the eyes usually mean anemia. It is important to be very observant if you notice this and do all you can to encourage her to get nutrients, especially copper, which will help her to absorb iron. She may be lethargic and off her food.

If the transition from normal goat behaviour to anemic is very sudden, it could be a bloodsucking worm such as barber pole worm. These are more common in warmer temperatures, especially in the spring, as goats are under strain from late pregnancy and kidding. Like all worms, these ones get into the goat when she eats food from off the ground of worm-infested areas, so can be prevented by following my instructions for feeding goats (page 38). Making sure that your goats have the right amounts of minerals as they approach the spring will also help prevent worm issues. The barber pole worm can stay dormant inside the goat until warm weather arrives, and then its attack is very sudden.

Intramuscular injections (see page 76) of vitamin B12 (1cc every 4 hours for a kid, double this for adults) can help, as can injections of 4cc of vitamin C for kids, double for adults. Feeding lots of kelp will help, as well as copper and diatomaceous earth (try to feed the diatomaceous earth at least an hour apart from the vitamin feeding).

Depending on how you feel about chemicals, and the certainty you have about whether it is the barber pole worm or not, you may want to resort to a standard chemical drench to quickly kill the worms before addressing the long-term deeper nutritional problems that have caused the worm troubles. If this is your choice, it's best to ask a vet for the right drench to use for this (and to get a test to confirm which worm it is), as the recommended drenches are changing all the time. Make sure to ask if there is any withholding period after drenching, where the milk should be avoided.

For natural remedies to deal with any kind of worm infestation, see pages 70-71.

Not ruminating (not chewing her cud)

Often when a goat is sick, she won't be digesting her food as well as she usually does. If you are observing her and she isn't eating much, or doesn't seem to be chewing her cud as often as she usually does, it helps to offer sodium bicarbonate (baking soda/bicarb soda) to her, as it will help to balance the acidity in her digestive system, something that the process of ruminating usually does for her. Dolomite has a similar effect and will provide extra minerals at the same time.

If a goat is recovering from being sick and isn't ruminating, it may be worthwhile to take some brought-up cud from another goat's mouth and feed it to the non-ruminating goat. This will transfer rumen bacteria that should increase in the sick goat's rumen to get her back to normal again.

Dehydration

If you want to encourage your goat to drink more, try adding some apple juice to her water. On hot days she will appreciate some ice in her water, and on cold days she will appreciate warm water.

Wounds

Hydrogen peroxide is good for first cleaning out the wound, diluted tea tree oil can also be used. Raw honey should then be applied to completely cover the wound; if it's a deep wound, put lots of raw honey inside it too. A dressing needs to then be made over it that the goat can't remove (easier said than done). For the dressing, you will need some sort of very clean gauze or other fabric that won't stick to the wound to cover the wound itself (and a bit extra around the edges of it), and then a stretchy bandage can be wrapped around her to secure the gauze; or gaffer tape (duct tape) can work either by itself, or in addition to the stretchy bandage. Keep in mind the way that goats move when you're doing the dressing, so that her normal movement won't make the bandage fall off right away. Check on her twice a day to make sure that the dressing is still on there, and change it after a couple of days, or sooner if there seems to be anything going wrong with it, adding more raw honey when you do so.

Skin and hair problems
Extra sulphur and kelp added to the diet will help. Irritated patches of skin or hair loss could also be from external parasites, so dust her with some diatomaceous earth if you suspect that this may be part of the problem.

Cure-all remedies

Probiotics
Probiotics will generally help a recovering goat, and one caught in the early stages of many illnesses. You can either give this to the goats as powder or tablets, or by adding whey with live cultures (such as the leftover whey from chévre or yoghurt) to her water. Raw apple cider vinegar added to the water also provides the benefits of probiotics.

Raw apple cider vinegar
Apple cider vinegar should be labelled 'unpasteurised' or 'raw'. This provides potassium, probiotics, and is a good general supplement for optimum health.

Vitamin C
This cures many issues, and can even be used to treat snakebite. The kind of tablets that are sold in the pharmacy are fine for feeding to the goats, but if you can find the sodium ascorbate form of vitamin C, you can feed it, as well as using it for injections (see page 76).

Garlic
Garlic will help to boost immunity, and help get rid of worms, but it should be used carefully with dairy goats, as it can make the milk taste like garlic.

Extra minerals

Feeding extra dolomite can help with a variety of ailments and stresses. Copper can be useful for many issues, but it's best to be careful about the amount given, especially for light-coloured goats; if too much builds up in the liver it can be toxic.

Rumen transplants

See "not ruminating", on page 74 for instructions. This is a quick and simple procedure that can cure a lot of digestive issues and minor 'goat cold' problems. I have seen long-term improvements in the health of goats that were often sick just by doing this.

How to give an injection

Firstly, the goat needs to be secured. Milking stands are good for this, or you can secure her against a fence, and hold her still (easier to do if you have a helper).

Anything that can be given intramuscularly can also be given subcutaneously (below the skin). Subcutaneous injections are easier to give, and less painful for the goat. To give a subcutaneous injection, pinch a loose bit of skin, either from her brisket (chest) or from the shoulder around where her front leg meets her body, and pull the loose skin outwards. Insert the needle into where the pulled-out skin meets the rest of the goat's body, keeping in mind that you are trying to inject just below the surface of the skin, not into it, and not too deep either.

For an intramuscular injection, the needle needs to go deep into the muscle (sometimes all the way up the needle). There are large, long muscles on the sides of the goat's neck that can be used for this, and also a muscle in the thigh. If you want to get prepared for doing this in case of a snakebite or other emergency, you can practise with a small dose of B vitamins or sterile saline.

The correct dose of sodium ascorbate for snakebite and complications from tick bites is five grams (10cc) intramuscularly every 2 hours, until symptoms improve.

How to check
eye membrane colour

Checking a goat's eye membrane colour is something that can easily be done while she's finishing her food after being milked. It's more important to check during the warmer months, when the barber pole worm is more active, and also during pregnancy, as she approaches kidding time and becomes more vulnerable to parasites.

Working carefully and gently through this process, first push the goat's upper eyelid down over her eyeball. Next you will need to gently push the eyeball down while using your other hand to gently pull the bottom eyelid down until you can see around a centimetre (half an inch) of eye membranes exposed.

To get an idea about what the different eye membrane colours look like, search online for 'FAMACHA', and you will find score cards that you can memorise the colours of, or print out and bring out to the goats with you. Red and deeper pink colours are good, pale colours indicate that there are worm issues, or nutritional deficiencies that need to be corrected.

Eye membrane checks are something that takes less than a minute to do, and it will mean that you will recognise problems early on and be able to respond to them before they become bigger problems, and will also be able to observe any sudden anemia.

Breeding

In order for there to be milk, your goat will have to give birth. After she's had a kid, she can often continue being milked until her supply dries up. The amount of time this takes will depend on the goat. If she's from a dairy breed known for long lactations, such as a British Alpine, she could give milk for her whole life without needing to be in kid again. Most goats can easily go for two years, so this gives the option of a continuous supply of milk from two goats, kidded in alternate years.

Goats are generally seasonal breeders. Bucks from European breeds will only be interested in breeding in autumn and winter. Does from the same breeds will have a heat (estrus) cycle of roughly 21 days during that time, so that she is in heat for a continuous 2 to 48 hours (usually somewhere in the middle of this), once every 21 days. The African breeds of goats (Anglo-Nubian, Nigerian Dwarf) can sometimes breed at any time in the year. The exact months of your goat breeding season will depend on your climate. If it gets hot in spring where you live, your breeding season might start in the last month of summer and end before the winter solstice. Cooler climates can have a breeding season that doesn't begin until around the autumn equinox, and lasts until a month or so after the winter solstice.

The age that a goat should first become pregnant will depend on how she is growing. If she is approaching the same size as the fully-grown does then it's fine to have her become pregnant a little earlier. For some exceptional does that are growing very quickly, there is no problem with taking her to the buck at seven months old, so that she's having her first kid when she's twelve months old. For most goats, though, it's best to get them in kid when they're around nineteen months old, so that they are giving birth around the time of their second birthday.

Because goats are seasonal breeders you will sometimes need to decide whether it's better to get your doe pregnant this season, when she may be slightly too young, or to wait until next season. The risks of having her get pregnant too young are that her body won't handle the pregnancy well and she'll be more prone to parasites, her growth may be stunted, and her udder capacity may not be as plentiful as it would be had you waited. The risk of breeding too late is that she may become too overweight to be fertile, but this is more easily controlled by restricting 'treat' feed and careful observation.

Signs of heat in a doe include excessive wagging of the tail, her bleating being louder, more insistent, or more frequent than usual. When you present a cloth that's been wiped on a buck during the breeding season, as she sniffs it she will get a love-struck look on her face and stand still while wagging her tail. Sometimes there will be a discharge from the vulva. If you want her to get pregnant, as soon as you notice she's in heat it's a good idea to either get her to the buck straight away, or to note the date down and organise to get her to the buck when she's next in heat. After your goat has been serviced by the buck, keep an eye out for the next estrus cycle. If she shows signs of estrus roughly 21 days after her last one then she may need to be serviced again, although sometimes does will show signs of estrus for one cycle after they've been serviced, so if you don't mind waiting another three weeks, keep an eye out to see if it happens a second time.

Sometimes a doe will go on heat as soon as she smells a buck, so if you're not sure when she is in heat you can probably get her pregnant just by putting her with a buck for a few days.

How you get a doe to the buck will depend on your situation. Here are some ways to do this:

Keeping a buck
If you have enough space in the yard, and can afford to construct an enclosure with very strong and secure fencing, you could consider keeping a buck. It's usually recommended to keep him separate from the goats that you're milking, as his presence will give their milk a 'goatier' taste, something you might appreciate in cheeses, but may not like in the milk. I haven't noticed this taste in our goats milk when we've kept a buck.

A buck will generally need a companion. A wether or a dry doe can do this. When you notice that your doe is in heat, walk her over to the buck and leave her with him until she's definitely been serviced. Whether you want to keep a buck will depend on a few things. If there are no bucks within a reasonable distance, and artificial insemination isn't available where you live, then your only option to get your does serviced will be to own a buck.

Bucks can be difficult to handle during the breeding season. If you live close to other houses, your neighbours might not appreciate the strong smell they have during the breeding season - they wee on their beards to make themselves more attractive to females (the bucks that is, not your neighbours I hope).

The personalities of bucks vary from buck to buck. Some can be easier to handle than does, others can be aggressive in the breeding season. Generally bucks that have been bottle-raised, or raised with lots of human contact will be easier to handle.

To keep a buck you'll need very strong fencing. During the breeding season he may run up to the fence with his horns, butting it with his full force. If he has another buck in with him, they may fight violently during the breeding season, putting more strain on the fence. The feeding needs of bucks are similar to dry does; grain is not recommended, but it's important keep a good eye on him during the breeding season, and if he appears to be losing weight or looking unhealthy, consider giving him some of the treats that you feed your does at milking time.

If you can't find a suitable buck, it's easy with a bit of planning ahead to buy a young buck kid and raise him to maturity yourself. If you can get one that was born fairly early in the kidding season, then he will be ready to service your does when he is around 6-8 months old. You can either bottle feed him, or see if your does want to give him their milk. If he goes in with your does to be fed by them it's important to visit him a lot, lead him around as often as you can, and generally make a fuss over him. The human contact he has as a kid will impact how easy he is to handle when he's grown. Raising a buck in this way is much more neighbour-friendly than bringing in a fully grown buck, because the little buck will be cute, and he won't be as smelly as a full grown buck.

Taking your doe to the buck

Unless the buck lives a short distance away, it makes sense to organise ahead of time with the buck's owner to make sure that he will be ready when your doe is in heat. Once you notice she's in heat, bring her there, and wait around to make sure she's been seen to. It shouldn't take long.

You could leave her there for a few days, but this involves another trip out, and if you're milking her she is likely to dry up if she goes for too long without being milked. Buck owners are generally understanding about this, but it's a good idea to keep an eye on your doe's cycles, and to let the owner know with plenty of warning when the next cycle is going to be.

Borrowing a buck

This is sometimes possible, but will depend on the buck's temperament. Some bucks are very difficult to handle in the breeding season and it's not a good idea to move them, whereas others are easily moved to your does. If you can find a buck you can borrow in this way it is often the easiest way to go, as you can keep him there for enough time to make sure all your does are pregnant, while keeping them in milk. It's best to try and keep him there for at least one full estrus cycle (three to four weeks). Two cycles are even better.

Artificial insemination

This is possible in some locations, but it means your goat misses out on the natural way of getting pregnant, and is instead serviced in a less 'romantic' way, by modern technology. On the positive side, it's easier to get her pregnant to a purebred buck that's free from diseases this way, and is sometimes less hassle than taking your goat on a drive to see the buck.

Pregnancy and kidding

Pregnancy

A goat's pregnancy usually lasts for 150 days from when she was serviced. Twins are very common in goats, single kids and triplets are often born as well. Your goat won't get very big until the last month of pregnancy, when the kids get noticeably bigger. A few days before she gives birth, her udder will fill up.

Pregnant does should be fed some grain or concentrated feed, and given access to kelp whenever they want. If they don't seem interested in kelp by itself, try sprinkling one or two teaspoons of it over their grain every day. This is a good way to get them used to the milking stand and being milked, as you can bring the doe to the stand every day at roughly the same time as you would if you were milking her, and stay with her while she eats, patting her. Once she is used to you, you can try touching near her udder, but most pregnant goats will not want you directly touching their udders. Once she is relaxed and happy for you to touch near her udder, try to do it often so that she gets used to it and is ready to be milked.

To give her the extra potassium she needs for an easy birth, adding some apple cider vinegar to her water or feed during the last month or two of pregnancy is recommended. I can't stress enough how important kelp can be for a pregnant goat. Kelp supplies a variety of micronutrients in a natural whole food form, and provides the easily accessed iodine that balances the intake of all the other vitamins. It's best to offer it free choice to goats, so they can take as much as they need without taking anything else with it. You may find if they've never had kelp before that they eat huge amounts of it at first, then once they've absorbed enough to correct any past deficiency, they will only eat small amounts.

If your doe is being milked while she is pregnant, it's best to give her a break for six to eight weeks before she gives birth. For these last weeks of pregnancy it is better for your goat to put all her energy into growing the baby rather than producing milk, and to give her udder a chance to recharge before her kid is born, to make sure she is ready to produce the large amounts of milk that she needs to provide in early lactation.

Drying a doe's milk up during pregnancy, if she's only being milked once a day, should just be a matter of ceasing to milk her and observing her to make sure she isn't in pain from it. If she's being milked twice a day, you will need to reduce to milking once a day before you dry her off completely. To help reduce her production you can also reduce her grain ration a little.

A heavy milker may have trouble drying off. You can try and gradually dry her off by taking a little less milk each day, or in some cases she may only physically be able to be dried off for a week or two before birth, or not at all. Your doe may bleat a bit as she notices that her udder is full, but she shouldn't be in pain, and her udder should not feel hot.

Get used to noticing the correct temperature for an udder before you begin drying her off, so that you'll easily be able to notice if her udder is starting to feel hot (a sign of a congested udder or mastitis). If your doe develops mastitis now she may be prone to it in the future, so be very careful and avoid this happening. Feeding her a teaspoon per day of dolomite while you're drying her off will help to avoid mastitis.

Sometimes it's hard to tell whether a doe is pregnant or not, and if you're worried that she isn't pregnant, and that drying her off will mean no milk for a whole year, then it's not the end of the world if you don't dry her off.

As the time of birth draws near, make sure her surroundings are clean by either getting rid of the old bedding and adding lots of extra straw, or by moving their paddock a few days before the kids are expected, providing them with some straw if you're doing this as well. As your doe approaches the time of birth, she may appear restless, inspecting her surroundings and deciding where she wants to give birth; she may make a 'nest' with straw, using her hooves to fluff it up, to make it into a comfortable and warm place to give birth. Carefully add some more straw or other bedding now if you think there isn't enough, being careful not to disturb her.

The birth

Well before the birth, you'll need to decide whether you will get a vet or farmer involved early if you see any trouble, or if there are some things that you feel confident enough to correct on your own, and at what point you would call someone else in. Decisions need to be made quickly, and it's best if you know what your plan is ahead of time and have contact details on hand so that you can quickly call someone in if you need to. Discuss your decision with your helper of choice and make sure they will be available around the time of the birth if they're needed. Arrange a plan B if you are worried.

It's good to have long, clean rubber gloves on hand, in case you or your helper need to assist the doe by rearranging the kids or pulling them out.

Despite the above warnings about being prepared, for the vast majority of goat births it's best to be hands-off whenever possible. Most of the time, your goat will easily give birth to a healthy kid or three. She might bleat a bit, and sound as though she's in pain at times, but it's completely natural, and all she needs is to be left alone in a setting she's comfortable with and observed quietly. With that said, it's a good idea to keep an eye on how much time has passed, if it's been more than an hour and a half since your goat began passing bloody mucus and she hasn't yet given birth to a kid, she might need help. If you think she needs help, it's best to have an experienced person such as someone that raises other ruminants such as goats, sheep and cows, or a vet to help you either on the phone or in person to begin with.

Always keep an eye on the time, not so much because a doe should be going through each stage of labour according to the clock, but because your own ideas and emotions may become clouded at the time, and it may look as though the kid isn't coming out the way it should, but usually nature fixes this without any help from us, and you may find that the time you've been watching and worrying hasn't actually been that long at all, and that things can be corrected without any assistance from you.

Sometimes the birth is stalled because the goat is not comfortable in her surroundings. If other goats are pestering her you may have to move the other goats somewhere else for the birth (in the backyard situation temporarily tethering the other goats is probably the best solution). Once you've moved the goats, try to leave her alone for a bit longer and see if the birth progresses naturally; usually this is all she will need, and it's best to allow her a bit of extra time to give birth naturally rather than intervening according to how much time has passed.

If you can't contact anyone, and think you need to help a kid out yourself, first check that the kid legs sticking out of the doe belong to the same kid, and are both from either the front or the back, not one from each end. You can do this by carefully feeling up the length of the leg until you reach the body, and then feeling for the leg next to it and making your way down towards the feet. While you're doing this, also check that there aren't any parts of the second kid that will stop the head of the first kid from coming out. If the legs sticking out are from different kids or one is from the front and the other from the back you may need to push one leg back in. While you're pushing this leg back in, gently position the other kid's legs so that they will easily come out together. Once you've positioned the first kid in this way you may want to wait and see if it can now be born naturally, or you may worry that the doe or kids are stressed and want to intervene to make the rest of the birth as quick as possible by pulling the kids out.

If you have to pull a kid out, try to pull as the goat is pushing, stopping when she stops, starting again as soon as you notice she's having another contraction. Carefully pull the kid out with two legs at a time.

When the kid is being born with it's back legs first, the umbilical cord will sometimes rupture during the birth, if the cord ruptures, then the kid will need to be helped out for the last stretch of it.

If the kid is born front feet first, but has its head turned back, you may need to reposition the head to be facing forwards. It's best to do this while the doe is having a contraction.

On many farms the goats will just give birth without anyone noticing until there is a lovely fluffy kid next to its mum. It's important not to be stressed around the goat during the birth, and to observe her - if your presence seems to make her uncomfortable, try to move away quietly until you're at a distance where she doesn't mind you being there. Don't get closer unless she needs help, and don't do flash photography or make any loud sudden noises; avoid anything that might distract her, even normal camera noise.

Does that are giving birth for the first time may need to be observed more carefully than experienced does. First time mothers will sometimes give birth and stare off into space, recovering from the pain or shock, not knowing that they have babies to look after. This is more likely to happen if your doe is not able to follow her instincts, so try to make sure she is comfortable with her surroundings and has a quiet and safe place to give birth in.

After the birth

Your doe will probably clean her kids' umbilical cords by chewing and licking them; even if she doesn't do this, the kids will probably be fine as long as they're in reasonably hygienic conditions, with plenty of fresh straw, and not much goat poo around. If you're concerned about infection, dip their cord stumps in some iodine, alcohol, or diluted tea tree oil.

Your doe will give birth to the placenta shortly after the kids have been born, and will probably eat it. Placenta eating has its advantages, giving the doe lots of nutrients at a critical time. If she doesn't eat her placenta it is nothing to worry about, just make sure to bury it or feed it to chickens within a few hours.

Your doe will probably lick her kids clean and dry, but if she doesn't and the weather is cold you might need to dry them yourself with a towel.

It's very important at this stage not to disturb the mother and babies unless absolutely necessary. If a goat is left to her own instincts, she will usually manage well as a mother, nudging the kids to their feet and giving them their first drink of colostrum.

Your goat will generally know what she's doing. She follows her instincts, and it's better not to intervene unless it's absolutely needed. The kids will generally start drinking their mother's milk fairly soon. Keep an eye on them, and make sure they have all had a drink within an hour of the birth.

It is very important that the kids drink colostrum within an hour of the birth. The longer they go without their first feed, the weaker they will become, and if left for too long they will be too weak to drink on their own, and too weak to keep warm.

If you observe the kids attempting to drink from their mother, and her moving away or otherwise not letting them drink, it's time to intervene by holding her still. She will sometimes need to be held still, or put in the milking stand, for the first few feeds, and later she will adjust to feeding on her own.

If the mother is moving constantly and not allowing them to even find the udder, you can tell that they want to drink by observing their behaviour. If they seem determined to be looking for something, it's the udder they're looking for and you will need to hold the doe still for them.

Sometimes a kid doesn't know what it's doing either and needs to be moved to the teats, and sometimes started off on the colostrum by opening the kid's mouth, squeezing colostrum into it, and then putting the teat in the mouth. The kid should start drinking on its own once it realises where the milk comes from, and will easily find the teat on its own next time.

If your doe has given birth to just one kid, the kid may have a favourite side of the udder. To avoid mastitis and congested udders, inspect her udder when she's eating her treats on the milking stand. If one side feels very full, then you may need to milk that side every day. If you're unsure you can always just milk a smaller amount from the udder, and then there will still be plenty of milk left for the kid.

If you arrive to find the kids born and already licked dry, observe them for a while. The mother has hopefully already formed a bond with the kids at this point, so it may be worthwhile to bring them to her udder, holding her still if you need to, and to make sure that they get some colostrum. Things can deteriorate fairly quickly if the kids don't feed within a few hours of their birth, it's always best to make sure the kids have fed within an hour of being born.

In the case of weak kids and floppy legs
If your doe has been fed plenty of kelp during the pregnancy, the kids have not arrived early, and the doe is in good health at the time of the birth, then the kids will probably be fine.

Sometimes there are weak kids who can not maintain a warm body temperature. They will feel cold to the touch and won't have the strength to move, and won't show much interest in milk. If their mothers are not snuggling up to them at all times they will need to be wrapped in blankets with hot water bottles, or to be put in the warming oven of a combustion stove with the door ajar. Woollen jumpers designed for lambs can also be useful, as these allow the kid to move around while still having an extra layer to help keep warm.

To tell whether a kid needs your help, observe him shortly after the birth. The kid should be able to stand up on his own after a few minutes. If the doe doesn't nudge the kid to stand, then you can gently lift him to his feet and see if he has the strength to stand without toppling over from wobbly legs straight away. Try and wait until after the doe has licked the kid clean before you try this. Usually the kids won't stand until after they're dry.

If a kid is unable to stand but still seems hungry and healthy, you can try making leg splints for him by getting some sticks and bandages, straightening each leg, putting a stick along it, and wrapping the leg and stick up in a bandage to keep it straight. After a few days the kid should have built up the strength to stand on his own.

Raising kids

It's much easier to raise kids with their mothers than it is to raise them yourself. Sometimes a doe will reject a kid and you will have to look after it, but the vast majority of times they are good mothers.

The only real disadvantage to keeping kids with their mothers is that you'll get less milk, but during this time a doe is giving so much milk that sharing it by keeping the kid separated at night and milking the doe in the morning she is still giving as much (or more) than you will get later in lactation milking her once a day.

You can begin separating the kids from the doe when they are around two weeks old.

You can keep your goat separated from the kids either by keeping them separated at night and milking in the morning, or if there is always someone at home you can keep your goat tethered away from the kids during the day, and milk her in the evening.

Another option for kids is to just keep them until weaning age (two to three months), and then sell them, and after this time you can have all the milk. This causes a bit of delay with getting the milk, and if your doe was kidding every year it may not be a good strategy, but if your goat is lactating for two years or longer after each kidding then 2-3 months off after the birth is not such a big deal. While the kids are with their mothers you can still get a cup or two a day out of the doe as you feed her treats on the milking stand.

However your goat kids are raised, make sure they have plenty of hay and water around from one month onwards, they will start to nibble at the hay, increasing the amount that they eat until they are ready to wean.

If you want the kids to be future dairy does, bucks, or pets, try to spend time with them often to get them used to the presence of people, let them climb all over you, and get them used to being led around by the collar or horn once they're big enough to be interested in solid food by holding a bucket of feed in front of them and gently leading them around as you move the bucket.

Castration

For buck kids that will be raised for meat or as pets, it's best to castrate them early, before they are a month old. The earlier it is done, the easier it will be, you may even be able to do it alone when the goat is very young, but by 4 weeks you'll probably need help to keep him still.

An elastrator, a tool which slips a tight rubber band around the balls to cut off the circulation, is probably the best option for a backyard goatherd. Elastrators designed for lambs will work for goats.

To use an elastrator on a buck, lie him on his back, making sure he is still, place the elastrator around the scrotum, as close to his belly as possible, so all of the testicles will be cut off from the rest of his body, and then release the rubber band so that it cuts off the circulation. Observe him closely to make sure both testicles are completely separated from his body. He won't like this. He won't like any method of castration, it's not a pleasant task for anyone involved, but it is usually difficult to sell a buck kid that hasn't been castrated, keeping a full buck as a pet is not a good idea, and the taste of buck meat is not something many people enjoy.

In the days following the castration, observe him to make sure that the band has stayed in place, and that he's not in a lot of pain. He may be less trusting of you and his back may appear slightly hunched up from pain to begin with, but after he's over the initial shock, he should be running and jumping around, drinking milk, and doing all the usual goat kid things. If the elastrator was used correctly, his balls should wither away in time from the lack of circulation.

Milking

How to milk a goat

Before you begin milking, make sure you have everything ready that you need for the milking and straining. The straining cloth should be boiled, the jars and funnel sterilised (see the next section for information about doing this). Fill a food bowl with your treat feed and place it in the feed bucket area of the milking stand.

The amount of treat feed to give a goat will depend on how much milk she gives. Some goats will handle higher amounts of grain than others, and will give more milk if you feed more grain, while other goats seem to do better with less grain. Around two cups of grain is a good amount to start with. Have your milking bucket close by, but not anywhere that the goat can easily knock over, or anywhere else that it can easily be knocked over or contaminated. I keep a small table near the milking stand, for the bucket and jars.

Walk the goat up to the milking stand and secure the head gate around her neck once her head is through. You will now need to clean her udder. You can either brush her udder with a dry cloth or the back of your hand to remove stray hairs and dirt, or if she's very messy, you can wash her udder with a wet cloth, and then rub it over with a very dry towel, making sure she is fully dry, as you're far more likely to get sick from dirty water dripping into the milking bucket then you are from a few stray hairs or bits of dirt.

If you're washing and drying the udder, you will need to use separate washing and drying cloths for each goat. Not many goat books advise following the first method (my preferred one) for udder cleaning, and prefer the more thorough washing and drying approach. A lot of the belief in washing goat udders comes from milking cows, where cows seem to be attracted to the muddiest part of the paddock and can have very dirty udders, goats on the other hand will find the driest place possible, and don't seem to get dirty very often. The simple method works for my family, as we keep our goats clean and dry with lots of straw, but if your goats are covered in muck, washing and drying is the best option. My mention of the simple method might be a controversial approach, but it is far easier to brush the udder quickly than it is to wash and dry the udder. If you're in doubt, or can't chill the milk quickly, or are sensitive to food contamination, wash and dry the udder, just make sure to dry it thoroughly.

Make sure you are seated in reasonable comfort. I sit on the edge of the milking stand, but plenty of people use stools instead, you should be able to sit there milking the goat without needing to bend your back and without stretching your arms out awkwardly to reach the teats.

Take two squirts of milk from each teat, milk it onto the milking stand, or onto the ground, or into a separate dish. By discarding the first squirts of milk from each teat we are decreasing the risk of contamination from anything that may have been lurking at the end of the teats. Once you have discarded these first squirts, and your goat's udder is clean, place the milking bucket close to the udder and begin milking into it.

To milk a full-sized goat, first place your thumb and index finger around the top of the teat where it meets the udder and close it off, then close your middle and ring fingers (or just the middle finger if her teats are small) around the teat to squeeze the milk out of it. Repeat this with one hand after the other, until it becomes more difficult to get the milk out of the teat. Remove the bucket, and then massage her udder or mimic the action that a kid uses on it by pushing against it with your hand, then place the bucket back under her and continue milking as you were before. This helps her to let down as much milk as possible.

If your hands get tired using that method of milking, you can alternate with another method where the thumb is not used, and just the index finger is used to close off the top of the udder. This method uses different muscles to the usual way of milking, but can't usually be done until the udder has emptied a bit.

To get the last of the milk from the udder, 'strip' the udder by using both hands on one half of it at a time, gently squeezing milk from that half of the udder into the teat, and then out into the bucket.

When it's time to stop milking, she will be giving the tiniest amount (or none at all).

It's easiest to watch someone else milking to learn, if you don't have anyone nearby try searching for videos online. To get your hands used to the action of milking, go through the motions of milking on your thumb. It's a good idea to learn to milk while the kids are still drinking their mothers milk, because the kids can help drink the rest of the milk, and your doe is not at risk of udder problems or drying up as long as someone is taking the milk. Once you are in a good milking routine with your goat, milking will take less time than the straining and cleaning. It takes me around five minutes to milk a goat with a good udder, a bit longer if her milk is slower to flow.

When you've finished milking your goat, leave her on the stand to finish her meal while you strain the milk. This extra time on the milking stand helps the teat to close before any bad bacteria can get into it. To strain the milk, place a funnel over the top of a glass jar, then cover the funnel with a sterilised thin cloth, such as butter muslin or cheesecloth, and pour in enough milk to fill the funnel. If it's taking a long time to go through the cloth, you may need to find a thinner, or more loosely woven cloth for next time, or you can gather the edges of the cloth in your hands and tilt them around carefully to try and get more milk through. The creamier the milk, the longer it will take to strain.

A milking routine

1. Boil the straining cloth (or boil it the night before)
2. Assemble everything you need during and after milking (e.g. clean jars and funnel on the table, clean milk bucket near the milking stand)
3. Bring the goat to the stand, clean her udder, milk her
4. Strain the milk
5. Take the goat back to her paddock
6. Repeat for other goats
7. When you've finished milking all the goats, wash the straining cloth and hang it up to dry, sterilise the bucket and funnel

Milking equipment and hygiene

When the milk is in the udder it is sterile, it is a perfect food for baby goats to drink, and you don't often hear of baby goats getting sick from their mother's raw milk in the same way that the media sensationalises cases of humans getting sick after drinking raw milk. Because of its neutral acidity, any milk, whether it is raw or pasteurised, is an easy medium for bacteria to grow in once it is out of the udder. This can either be beneficial, when we encourage good bacteria in order to make cheese, yoghurt and other fermented foods, or it can be very bad in some cases. If you're drinking the milk right away, hygiene is not much of an issue, but the longer you wish to store it, the more careful you need to be about anything that will be in contact with the milk, and the storage conditions that you keep it in.

To prevent contamination by the wrong bacteria on milking equipment there are two main approaches, the most sustainable and healthy option being heat sterilisation, which involves heating anything that the milk will be in contact with to a safe temperature before the milk touches it. The other main method is with chemicals.

After the equipment has been sterilised, it needs to be kept away from anything that could contaminate it. Milking buckets with lids are recommended because you can then keep the inside of the bucket and the funnel away from any contaminants. I use a bucket with around seven litres (seven quarts) of capacity, but around half this size is sufficient, because you will want to empty the bucket as soon as you've milked each goat to avoid the frustration of the bucket being kicked or stepped in with all your milk in it. Buckets larger than seven litres will be difficult to fit underneath the goat, so search for a small stainless steel bucket with a lid – it can be tricky to find, but it's something that only needs to be purchased once, so it's worthwhile getting the right one to begin with. A stainless steel stockpot can be used instead of a bucket; try to find one without air holes in the lid and you will have something that works just as well as a bucket with lid.

Heat sterilisation can either be done with boiling water, or in the oven. The use of boiling water can result in injuries and rude words, so the oven is preferable in some ways. To use the oven, everything you put in it should be either metal or glass. Plastic can go in to dry after being boiled once the oven has been turned off, but anything put in the oven for long enough to sterilise it should be able to withstand the heat, so I recommend stainless steel milking buckets, stainless steel funnels, and glass jars.

To sterilise using the oven, place your clean bucket and jars in it upside down, and switch the oven on to 110°C (230°F). Don't use ovens with visible (orange-glowing) electric elements for sterilising glass, as the glass can crack. Leave the oven on to heat up, and once it's been fully hot for at least five or ten minutes, test it by touching something that is sterilising in there – it should be very hot to the touch. Turn the oven off, close the door, and leave it to cool down. For jar lids and plastic funnels, first submerge them in boiling water for 30 seconds, then drain, and place in the oven as it cools down, to dry them and keep them sterile until you need to use them. Be careful that your plastic will stand up to this, some plastics are flimsier than others, but the plastic Ball Mason jar lids stand up to this without problems, as do good funnels designed for jam.

For the straining cloth, I rinse it and hang it out to dry after every milking, and then boil it when we're ready to begin milking next.

If you need to sterilise everything with boiling water, boil enough water so that you have plenty to pour over everything that needs to be sterilised. You are trying to heat up the surfaces of the milking bucket, funnel and jars to kill any nasties that might be lurking there, so slowly pour plenty of boiling water over all the surfaces.

The jars should be wet (but with no standing water) before you add boiling water, so that they don't break. Pour the water into the jars, filling around one third of the way, place the lids on, and carefully turn them on their sides, exposing every side to the boiling water until it is too hot on the outside to comfortably hold onto. These days I rely on a wood cooking stove that's often slow to boil water in the mornings so at night I put the straining cloth in the bucket with the funnel, and leave it all submerged in the boiling water overnight, draining it all in the morning.

Every time you empty a jar of milk, rinse it twice with a little cold or lukewarm water, and leave it until you're next ready to sterilise jars. Leaving jars sitting with a little bit of milk in them makes it very hard to remove it later.

Handling and storing milk

Once the milk is in the jars (if you're not drinking it right away) you should chill it as quickly as possible. I surround each jar of milk with ice-filled containers (I use the "ice bricks" made for eskies or coolers). You can even chill the jars before you begin milking, so that the cold jar begins chilling the milk as soon as it is strained. After the milk has been chilled, it's important to keep it at a low temperature, and to not let the temperature fluctuate. A fridge door, where many people think to keep their milk is not a good place for longer-term storage of milk, although there are no problems keeping milk you'll be using within the next day there. The temperature fluctuates too much in the fridge door, and it's also the warmest part of the fridge. It's best if the milk is on a shelf on the fridge, towards the back is best. Avoid leaving the milk sitting out at room temperature for entire meal times or similar periods of time unless you plan on drinking the rest of the jar very soon.

Don't place warm containers of leftovers or anything else next to the milk. If you're adding a lot of room temperature and warmer things to the fridge at once, add some ice-filled containers right next to the milk to make sure that the milk stays cold.

In general, the quicker you'll be drinking the milk, the more relaxed you can be about its storage. Bacteria need both the right temperature and the right amount of time to multiply to dangerous levels.

Now that I rely on a small off grid system without a fridge I find that in winter we drink all the milk within around 24 hours, so the temperature of an unheated room out of the sun ($10°C/50°F$) is cold enough for the milk, even without ice bricks. In summer there is a lot more milk to handle, and on the hottest days anything that doesn't get purposely fermented into cheese or yoghurt begins to ferment on its own within 24 hours unless I'm super careful about switching the ice bricks around.

1 litre (1 quart) mason jars and 650ml (22oz) passata bottles, are good sizes for storing fresh goats milk. If you want to make the easiest chévre from milk warm from the udder (see page 120 for the recipe) it is worth having a 2 litre (2 quart) jar for this purpose.

Dealing with excess milk

If you're lucky in your search for goats, and the goats you thought would only give one litre (one quart) a day each end up giving two litres or more, you may find you have more milk than you can use for drinking and cheesemaking. Depending on local laws, selling or giving away the excess milk to trusted people who aren't going to dob you into the nanny-state or brag about it on gourmet internet blogs is one option. A herdshare arrangement may also work. The legalities of selling raw milk varies depending on the country and state, so you may not end up having to jump through any hoops at all.

Another option is to feed the milk to animals. Pigs can drink the milk as it is. While chickens will drink milk, they will gobble it up in its solid form far more readily if you make 'chook ricotta' by following the ricotta recipe on page 102. Milk left sitting at room temperature for too long will often begin to turn to cheese on its own, and can also be fed to animals.

Goat's milk is great for the skin. You can either add it to baths, or dilute it and use it on your skin. Fresh goats milk applied to sunburn and other skin conditions is soothing and healing. Goat's milk soap can also be made.

Diluted raw milk sprayed onto pasture or garden plants works as a great anti-fungal and fertiliser. This is my way of making lemonade from lemons when a goat puts her foot in the bucket – at least the garden is getting a drink!

Excess whey from cheesemaking can be fed to pigs and chickens, and made into whey cheeses such as gjetost. Whey with live cultures from chévre and hard cheese (not acid-curdled whey from ricotta or paneer) is great for adding as a starter culture to fermented vegetables like sauerkraut and pickles. Whey can also be used in place of water in stock and bread, as soaking water for grains, and used as cooking water for vegetables, grains and pasta. Whey from cultured cheeses (not acid-curdled ones) can be diluted (one part whey to nine parts water) and used to water plants.

CHEESEMAKING AND RECIPES

The cheeses that will work best in your home will depend on what your goats are eating, the cheesemaking and maturing environment, how much time you have, and what you personally like to eat.

You may find yourself eating a lot more soft cheese than you previously did, because it's so easy to make and at its best when freshly made. If you have any cheesemaking failures along the way, the times when it does work will taste even better. Cheesemaking can be a complicated subject, yet to begin by making some simple fresh cheeses is really easy, so that is where we'll start.

An important note about ingredients:

Always use salt without any anti-caking agent, iodine, or other additives. For my recipes you will need a fine-grain salt, such as fine Celtic sea salt or fine Himalayan salt. If you can't find either of these unrefined salts you can use a refined fine salt such as flossy salt or dairy salt, or smash up some rock salt in a mortar and pestle. If you're using a refined salt, make sure to reduce the amount in these recipes by around ¼ to compensate for the increased saltiness.

For consistent results with renneted cheeses please make sure that your rennet has been stored in a dry environment, and is not past its best before date. At the time of writing this I recommend Walcoren, which is a certified organic animal rennet. If you can't find this, my next choice would be any other animal rennet in tablet or powder form. Vegetarian rennets are easier to find, but are often made by a GMO process (although it is debated whether the final product is actually GMO or not). Vegetarian microbial rennets are not genetically modified, but can give cheeses a bitter taste (there are some great cheeses made with this rennet though, so it might be a good choice for strict vegetarians). Liquid rennets are easier to measure, but have more additives than dry rennet.

Always find out first how strong the rennet you are using is before you begin. You may find that the 'regular' dose of rennet to use changes depending on which brand you are using.

Tomme, Asiago and the Alpine cheeses use a stronger dose of rennet to the regular dose, so once you find out the regular amount of rennet to use for Cheddar, Gouda, feta and so on, you can double this amount to make the harder cheeses.

Cheese press, 800g cheese mould with follower, dairy thermometer, slotted spoon, and cheesecloth. With these tools you can make any cheese, but all you need to get started with soft cheese is the cheesecloth.

Ricotta and Paneer:
Acid-curdled cheeses for beginners

A good introduction to cheesemaking is by making a whole milk ricotta or paneer. These can be made with ingredients you already have in your kitchen, and are a good way to use up milk that's been sitting in the fridge longer than you'd like to drink it raw. Ricotta can be used in cheesecakes along with savoury and sweet tarts. Paneer is ricotta that has been salted and pressed, to form a block of cheese that can be pan fried and then eaten by itself, used in curries, or in place of tofu in recipes.

A whole milk ricotta is basically milk that has been heated up to an almost-boiling temperature, and then the right amount of vinegar or lemon juice is stirred through and the mixture is left alone for five to ten minutes until it separates into curds and whey. The mixture is then strained through cloth, to separate the whey from the curds.

Ingredients
2 litres (2 quarts) milk
¼ cup vinegar or lemon juice, plus 2 tablespoons more if needed
Optional ½ teaspoon salt

Method
Slowly heat the milk in a pot over a medium-low heat, stirring every now and then to ensure the bottom doesn't scald, until it is very hot and almost boiling (it should reach a temperature of around 190°F/88°C). Continue to stir and slowly pour in the vinegar while stirring for a few seconds more. Stop stirring and leave it to rest - it should begin to separate into white curds and green-yellow whey within a minute. If the whey appears to still have a lot of milk in it at this point, add a tablespoon or two more of vinegar while stirring. Leave it to rest for 15 minutes.

While the milk is heating up, bring a small pot of water to the boil. Add a piece of butter muslin, cheesecloth, or another thin cloth and boil for 30 seconds, to sterilise it. Pour into a clean metal colander, pouring the hot water over the colander to sterilise it.

When the muslin has cooled, carefully arrange it to cover the colander, in one layer if you're using butter muslin, or in four layers if you're using loosely-woven cheesecloth. When the curds and whey have rested for 15 minutes, carefully pour them into the colander. Leave to drain for around 20 minutes. You now have ricotta.

If you're adding salt to the ricotta, mix it through now. Use half a teaspoon of salt if you want to make paneer.

To make your salted ricotta into paneer: loosely fold the muslin over the salted ricotta to form a solid and even parcel less than one inch thick (but not too thin), making sure there are no knots or lumps in the muslin that would cause dents in the cheese. Place this on a baking sheet, chopping board or flat dinner plate and cover it with another baking sheet, plate or chopping board, then add five pounds (2.5kg) of weight on top of that - this can easily be achieved with canning jars filled with water, or with some tins of food. Leave this to press overnight before using in recipes or storing for later.

Some more whey will come out of the curds during the pressing process, so be sure that your baking sheet has a rim around it to contain this, or place it on the draining board of the sink or somewhere else where it can drip.

Ricotta and paneer will keep for a week or two in the fridge.

To cook paneer, cut it into shapes, heat a frying pan over medium-high heat, add some ghee or lard once it's hot, then add the paneer, cook on one side until it's browned in places, and then flip over and cook the other side.

Yoghurt cheese: Another beginner recipe

Another good recipe to begin with is to simply get some natural yoghurt made without any thickeners (the ingredients should just be milk and cultures) or some of your own goat yoghurt (page 152). Greek yoghurt will yield more cheese and less whey than thinner yoghurts. Homemade goat yoghurt tends to be fairly thin and needs a bit more draining time in the colander.

Firstly, allow your yoghurt to sit at room temperature until the whey has separated from the curds. If you eat a spoon of yoghurt out of it, then if you have a look at the remaining yoghurt later in the day (or the next day), there should be a pool of light green liquid where the spoon was taken out - this is the whey. For thinner yoghurts, you will notice that the yoghurt when it is ready begins to stay together in a clump when the jar is tilted, sometimes you will see a tiny amount of whey when you do this.

Next, place a large sterilised piece of cheesecloth (see the directions in the ricotta recipe) or other thin cloth over a colander and drain the yoghurt over a bowl until no more whey drips from it, usually between twelve and twenty four hours.

For best results, tie the cheesecloth up from all four corners and hang it from a wooden spoon over a tall saucepan, making sure it won't be sitting in whey. This will extract more of the whey than using a colander, but it's also a bit harder to begin with, as some of the yoghurt can easily get squeezed out of the cloth when you're tying it. Once the cheese is drained, add salt to taste, then hang it back up again for a few hours to allow the salt to drain more of the whey out, or use it right away as a spread. It will get a bit saltier as the liquid drains out, so err on the side of too little salt over too much; you can always add more later.

If the curd formed is firm enough (you may need to hang it for a bit longer), you can then roll the cheese into balls and marinate it in some olive oil and herbs, or you can use it right away in cheesecakes and other recipes.

Yoghurt cheese (also known as labneh) has a very sharp and yoghurt-like taste, and is best complimented by other strong flavours, such as dukkah, za'atar, or other nut and spice mixtures. Soft labneh can be drizzled with olive oil, sprinkled with spices and served as a spread with vegetable sticks, roasted vegetables or flatbread. In sweet dishes it is complimented by honey.

The whey from yoghurt cheese can be used as for any cultured cheese whey, so making this cheese can be a quick way to get some whey ready for making lacto-fermented vegetables.

The whey and the cheese will keep for around two weeks in the fridge.

Farmers Cheese and Quark

Sometimes farmers cheese made from just raw milk and nothing else is too easy to make, other times it just isn't possible. In winter it can be hard to find a temperature warm enough for slowly culturing the milk, whereas in summer without a fridge, the milk will quickly begin to turn to quark.

To make this cheese, just set a jar of raw milk at a warm room temperature (preferably around 22°C (71°F) until it turns into curds and whey (around 48 hours). Strain through cheesecloth for a few hours, add salt if you like, and you have a cheese. For more consistent results, add a small amount of some sort of starter culture, such as viili, kefir or whey, at the rate of around 2 tablespoons starter to 1 litre milk, before culturing and straining as you would for raw milk.

Understanding the Cheesemaking Process

Making any kind of cultured and renneted cheese involves roughly the same ingredients and steps, with the addition of aging for hard and aged cheeses. The first step is the culturing of the milk.

If you just want to quickly go on to making your first chévre, feel free to skip this section where I explain the entire process. If you want to understand how and why it all works please read on!

Cleaning and preparing

Before making a cheese I think it's important to get all sourdough and kombucha (and anything else with yeast in it) away from where the cheese will be made, to stop the yeasts getting into the cheese. I keep these on the other side of the kitchen, and will either begin the cheese before I begin the bread (while keeping the lid on the cheese), or I will make sure the bread has been mixed a while ago, so that there aren't any stray bits of yeast or flour floating around after kneading.

It's important to have clean hands. Soap and water will do a good job.

Most cheesemakers like to sterilise anything that will touch the cheese. I do this by bringing my cheesemaking pot with some water in it to the boil, and then pouring it into my heatproof measuring jug that I use for diluting rennet, along with any spoons I use for getting cultures and salt, and spoons I use for mixing the cheese. Later on I will usually bring another pot of water to the boil for sterilising the cheesecloth, colander, cheese mould and follower. Sterilising anything oven-proof in an oven heated to around 110°C (230°F) is also an option.

Water for diluting the rennet with, or washing the curds in the case of Gouda and Havarti needs to also be prepared. If your water is chlorinated, standing it out overnight should get rid of the chlorine.

If your water might have bad bacteria in it, boiling it and then allowing it to cool to room temperature is a good idea.

Culture

Raw milk contains many beneficial cultures already, but to speed the process of cheesemaking up, and make the fermentation more reliable and less likely to be contaminated by unwanted bacteria, a starter is added. A starter can be fresh whey from a previously successful batch of cheese, room temperature yoghurt such as viili, milk kefir, or dehydrated starter cultures. You can also get away with using yoghurt from the shop if it has live cultures, usually this kind of yoghurt will work best in cooked curd cheese recipes and other recipes that are fermented at higher temperatures, as the cultures in there aren't very active at the lower temperatures used for most cheeses.

For the home cheese maker aiming towards sustainable cheesemaking, either viili yoghurt or milk kefir made from milk kefir grains is ideal, as it can be fed in small amounts in a similar way to a sourdough starter, and you're not obliged to use it right away for cheese, as is the case with whey. Viili yoghurt is a pleasantly sweet and easily made yoghurt, and kefir makes tasty and healthy smoothies, so it's often worth keeping at least one of these cultures on hand even if you don't want to make cheese. Follow my recipe for viili on page 152, and milk kefir on page 151 to make your own. For cheesemaking, the 'short' viili culture which produces a gelatinous yoghurt is a better choice than the 'long' viili which makes a stringy-textured yoghurt.

Cultures are at their best when freshly fed, so if you haven't fed your kefir or viili for a while, make sure to make a fresh batch around 12 to 24 hours before making cheese, and to taste it to make sure it has cultured correctly. Milk kefir is generally more resilient to being forgotten about than viili.

Many home cheese makers begin their journey by using dehydrated starter cultures. The benefit of these cultures is that they can just stay in the freezer and come out when you're ready to make cheese, so they need no feeding or extra care, but the downsides are that their packaging is wasteful, they are made in labs far away from where we live, using questionable methods, and their use cannot be sustained over the long term. It's not something like a sourdough starter, yoghurt, or milk kefir that can just be fed again and again from kitchen ingredients. When the packet runs out, more culture needs to be purchased.

Whatever your choice of culture, it is added to milk at the right temperature for the cheese recipe, usually a warm room temperature (I often use milk warm from the udder as soon as I finish the milking). It should not be cold to the touch, but too much heat will kill the cultures. Cheese is traditionally made in some areas by mixing the warm milk fresh from the udder with roughly the same amount of cold milk from the night before to create the ideal temperature for many cheeses. The temperature doesn't need to be exact, but a temperature that is a little colder than in the recipe will mean the culturing stage will take longer; a little warmer (as long as it's not so warm that it kills the culture) will result in faster culturing, and sometimes growing different cultures to the ones that grow at lower temperatures – if you are using villi, kefir or whey this is not an issue (these contain thermophilic cultures as well as the lower temperature mesophilic cultures) but if you're using freeze dried mesophilic culture too much heat will result in unreliable culturing. Too cold a temperature and the cheese will be very slow to culture, and the rennet won't be very active. You can make a cheese without a thermometer if you allow plenty of time and observation for the cheese to get to the right stages. "Baby bath temperature" can be found by touching the milk with the inside of your wrist: If it seems hot, then it's too hot, if it's pleasantly warm, or just slightly lukewarm, then it's fine to begin the cheese recipe.

Rennet

Once the cheese is cultured, rennet is added. Rennet is traditionally made from part of a stomach of a very young ruminant. It's possible to make this at home, however most home cheese makers buy their rennet in liquid or tablet form. Rennet tablets or powder made from animal rennet (usually calf rennet) are the purest option, but a number of liquid rennets, including vegetarian rennets, are available. Cheesemaking supply shops online have the best choice of rennets.

To add rennet to the cheese, it is first diluted in water, and then gently sprinkled over the cultured milk before being thoroughly mixed through for around two minutes in an up and down motion, to ensure that the rennet is evenly distributed throughout the cultured milk and not just stirred into the top. The cheese mixture is then left alone at the right temperature until the milk separates into curds and whey. The time this will take depends on the recipe, the temperature, and also the amount of rennet and culture added, along with how potent the rennet and culture being used are. Lightly cultured chévres and other soft cheeses are often left for twelve hours to form a soft curd, the rennet is often added at the same time as the culture in these recipes. Most cheese recipes use a higher dose of rennet and will have this stage lasting one to two hours.

It's important that once you have diluted your rennet in water it is added to the milk within half an hour, or the rennet will start to lose its effectiveness.

The rennetting stage is over when the curd achieves a 'clean break' from the whey. Put a very clean finger or knife diagonally into the curd for around an inch, then gently rip it up through the curd. The curd will still be fairly soft at this stage, but should be strong enough to hold an edge when tested in this way, rather than disintegrating into tiny pieces, and the whey should be separate from the curd. It should be clear, not milky. Leave it for a longer time if it doesn't break cleanly, and check that the temperature is still warm enough (rennet is most active at around 90°F/32°C). If it is taking a very long time, you may want to try gently stirring through some more diluted rennet, and allowing it another hour to set.

The amount of rennet to use can vary depending on which type you have, so please check your source of rennet first to find out how much is normally used. Most cheeses will use the normal amount (1/4 tablet or $1/32^{nd}$ of a teaspoon per gallon (4 litres) of milk for dry rennet, 1/4 teaspoon for liquid rennet), but very hard cheeses will use a bit more, and chévre will use a lot less.

Additional steps

After the rennet has set the curds, soft cheeses are often just placed in cheesecloth to drain and then salted. Most cheeses have additional steps, usually involving cutting and stirring the curds, to release whey, increase the acidity, and change the texture of the curds. These steps are usually what makes each cheese so unique.

Cutting the curds

Most cheeses, except the very soft ones, go through this stage. When the rennet has set the cheese, a long knife is often used to cut the curd roughly into cubes, the knife is first used vertically, to cut through the curds so that the surface is covered in squares, then the knife is used diagonally, to try and cut these square strips into cubes. It's best to let the curds rest for around a minute before beginning the diagonal cut, the curds are very fragile at this stage, and it helps them to not break into tiny pieces. The size of the curds will depend on the recipe; very hard cheeses will cut the curd into tiny pieces, sometimes using a whisk, while Brie-style cheeses will use curds the size of a golf ball.

Stirring the curds

Most cheeses that have had their curds cut will then be stirred to release more whey. The time this will take, the temperature, and how it will be handled will depend on the cheese being made. Some cheeses are "cooked curd" cheeses and are slowly heated up while being stirred to encourage thermophilic cultures. You don't need to stand there the entire time stirring, but it's good to be around in the kitchen during the stirring stage, and to stir as often as you can, especially if the cheese is being heated, because you don't want the curds at the bottom of the pot to overheat, and leaving the curds alone for too long without stirring will result in the curds clumping together and not expelling as much whey as they should. Stirring around once every five minutes works reasonably well.

Washing the curds

Some cheeses such as Havarti and gouda have an additional step of washing the curds to give the cheese a mild taste. This is done by removing a set amount of whey, and replacing it with warm water, while increasing the temperature and stirring the curds.

Cheddaring
This is only done for Cheddar-style cheeses. Real cheddaring involves draining the whey, allowing the curds to clump together, cutting the curds into slabs and resting the slabs on top of each other to begin pressing the cheese while it's still developing acidity. The slabs are moved and restacked often to release as much whey as possible.

Draining the whey
When the stirring has been completed, curds are allowed to settle to the bottom of the pot (this usually takes around five minutes), and most of the whey is then carefully poured off. The curds are then gently placed into cheesecloth to allow the rest of the whey to drain. Some cheeses are salted at this stage, others instead rely on the salting of the rind during the aging process.

To drain soft cheeses such as chévre and yoghurt cheese, the curd is first drained in a cheesecloth over a colander, and then the corners of the cheesecloth are tied together over a wooden spoon. The wooden spoon is suspended over a deep pot or bowl with enough room for whey to gather at the bottom without touching the curds.

For hard cheeses, and cheeses set in moulds, after the initial large amount of whey is drained in the colander, the cheese, sometimes in the cheesecloth, sometimes without, is then put into a cheese hoop or cheese basket for the whey to drain while the cheese settles into the preferred shape.

Shaping and pressing
After the whey has been drained off, the cheese is moved to the form that will allow the rest of the whey to drain away. This can be cheesecloth over a colander, in the case of soft cheeses, or it can be a cheese hoop or basket with a follower, sometimes with weights or a press to press on it, as is used for hard cheeses. The cheesecloth covered curds are moved into the mould, evenly arranged so that they're relatively flat on the surface, and the remainder of the cloth is carefully arranged in a way as to not leave big dents in the curd from where it's been twisted.

For hard cheeses, I usually use an 800g cheese mould. This mould is 12cm (4 and 3/4 inches) in diameter, and 13.5cm (5 and 1/4 inches) high.

Weights are added for pressed cheeses. If not much weight is needed, then canning jars filled with water can be a good option. Some cheesemakers use jars of warm whey, as the warmth of it helps to press more whey out of the cheese and achieve the same results as using heavier weights.

A cheese press is recommended for some hard cheeses. These come in a variety of styles, but generally all of them are using pressure generated through springs and similar devices to press down the follower on top of the cheese hoop to expel more whey. A cheese press that goes up to fifty pounds (twenty two kilograms) of weight, with the ability to use half of this amount of weight when needed is a good option for the home cheesemaker. It's not something you need right away, and many cheesemakers happily live without one.

Generally when a cheese is pressed, it will first be pressed at around half the final amount of weight, then the cheese will be removed from the press, flipped over and pressed with the higher amount of pressure for the recipe.

Salting

The time to use salt in cheesemaking will depend on what style you're making. For fresh cheeses it is often mixed though the draining or freshly drained curd, for most hard cheeses it is rubbed on the cheese after it has been pressed.

Larger-scale cheesemakers will often soak cheeses in a salty brine rather than directly salting the surface, but I think on a home-scale it is much easier to just use salt.

The type of salt to use is very important. It should be a pure salt with no additives such as iodine or anti-caking agent. Recommended salts for cheesemaking are Himalayan salt, Celtic sea salt, and any good sea salt without additives (keep in mind that some sea salts are refined and therefore saltier than others, so you may need to adjust the amount of salt in the recipe depending on what type you are using). Your salt will need to be finely ground. If in doubt, crush your salt with a pestle and mortar to make the grain of it finer.

The amount of unrefined salt to use when salting the surface of a hard cheese is roughly one tablespoon for every four litres (one gallon) of milk used in the recipe. Reduce this amount by 1/4 if you're using refined salt.

For soft cheeses, adjust the salt to your taste, keeping in mind that it will become saltier as it drains, and that you can always add more later on if you need to.

Ashing
Cheeses can have a coat of edible charcoal sprinkled over them, to form a contrast between the white inside and the charcoal layer, to add a subtle nutty flavour to the cheese, and to change the acidity of the cheese surface to encourage the bloomy white rinds to grow.

The ash is generally added to the cheese when it has almost finished drying in an airy place, so that doesn't have drips of whey coming out of it any more, but there is still some moisture on the surface for the ash to adhere to.

Ash can also be sprinkled in the middle of cheeses. This was generally done when there wasn't much milk, so the first half of the cheese was made on one day, sprinkled with ash to stop it forming a rind or getting bugs on it, and then the second half of the cheese was made another day and put on top of this before the cheese was pressed. This method could be a way for backyard goatherds to make larger cheeses when there is only a small amount of milk available every day.

Fresh chévre can be formed into a log, rolled in charcoal, and then eaten as-is. Aged chévres such as Valençay are often treated in the same way, to form a charcoal stripe in between the bloomy white rind and the cheese inside. Ashed Brie is another well-known ashed cheese where the charcoal adds an extra dimension of flavour that enhances an already great cheese.

Cheesemaking ash can be made at home, by selecting oak or apple tree twigs (or other non-poisonous twigs) or grape vines, keeping them all around the same size, and putting them inside a tin can with a lid or wrapping them in foil, punching a hole in the lid or package for air to escape, and then putting it on a fire until smoke comes out of the hole. The length of time this will take will depend on the size of the twigs. Once the charcoal is made it can be crushed in a mortar and pestle and used on your cheeses. Charcoal for cheeses can also be bought from cheesemaking supply shops, where it's generally known as 'vegetable ash'.

Airing

Hard cheeses are aired to evaporate any excess moisture and to develop a rind before they are moved to the cheese cave. The space for doing this should essentially be open to fresh air, but constructed in a way to stop insects landing on the cheese. The temperature for this doesn't have to be quite as cold or stable as the cheese cave, but it is best to keep it as low as possible, so try to keep it away from the cooking stove, the sun, and other sources of heat.

Some old cupboards have sections with fly screen on the sides that work really well for this. The type of meat safe made from fly screen or metal punched with holes, placed in a draughty place where it's not too warm will also do the job well. Fly-proof food covers, constructed as wire hoops covered with fly screen will also work. If there aren't any flies or mice around you can probably just air the cheese out in the open.

Aging cheeses

The aging time for a cheese will depend on the recipe. Soft cheeses are best used right away. Feta and halloumi need a few days in salt or brine before they're ready. Camembert and some other soft surface-ripened cheeses need four to six weeks. Hard cheeses generally need two months or more before they're ready to eat, and if they're kept in the right conditions their flavour will be better if they're left for longer. Some hard cheeses such as vintage Cheddar and parmesan are aged for eighteen months to two years. The optimum aging time for each cheese will depend on the conditions they are kept in, how active the cultures are, the size of the cheese, and the style of cheese that you're trying to make.

Cheeses should not be packed in close together when they're being aged. There should be space around each cheese to allow for airflow.

During the aging period, cheeses need to be carefully flipped over around once a week, and the rinds of some cheeses will need to be carefully rubbed or brushed, to discourage unwanted fungus and encourage a rind to form. You can also add extra salt to the rind at this stage, especially if the "cat hair" fungus is growing on your cheese (this is a harmless fungus that signifies that the cheese hasn't had enough salt - it is a fluffy grey fungus that grows to look a bit like clumps of cat hair). To get rid of it, rub salt on the places where you see it (or over the whole cheese).

The best surface for a cheese to age on is wood, but specialist plastic cheese aging mats also do the job. Sushi mats can also be used. Try to move your cheese to a dry area of the aging surface every time you check on it.

In general, cheeses are best aged at around 55°F (13°C). For cheeses sealed in wax, the humidity doesn't matter, but for natural rind cheeses, it's best if the humidity is generally around seventy to eighty percent. Traditionally, underground caves would provide the optimum environment for aging cheeses. For the home cheesemaker a number of options are available:

Seasonal cheese aging

For the home cheesemaker in a temperate or cold climate who doesn't want electrical gadgets and can't build an outdoor cheese cave or indoor cold cupboard, finding a place in the house that is consistently at the right cheese aging temperature for at least two months of the year and making cheeses during that time is the best option.

Most successfully for us has been an old meat safe in an unheated room with an open window that doesn't get direct sunlight. We have also successfully used an old fridge, switched off, on a shady porch. Crawlspaces under the house can work as well, as long as the cheese is on a side that doesn't get much direct sun (east, or either north in the northern hemisphere, or south in the southern hemisphere), if you're using a crawlspace, the cheese will generally need to be kept in something that is rodent-proof, such as an old metal meat safe, or a container with a vent on the lid.

With this method, aged cheeses are usually made in autumn and winter, when the temperature is right for aging. Observe your space throughout the year, and notice if the temperature changes much on hot days; you may find that you can age cheese at any time of the year this way.

Wine fridges, and other modified thermostat fridges

An option for making cheese at any time of the year is to use a wine fridge set at the right temperature, or a bar fridge with a thermostat gadget added to it to keep it at a higher temperature than the standard fridge temperature.

Passive cheese cave or root cellar

For anyone serious about cheese making who doesn't intend to move house any time soon, there is option of constructing a cave in your own yard. This could be an aboveground cave, made from straw bales or cob, or the traditional underground, or partly underground root cellar, often set into the side of a hill. To have the humidity that cheese aging needs, it's best for it to have an earth floor. If you're planning on aging a lot of cheese at once, it's a good idea to have a vent towards the top of the cave, to allow gases from the cheese aging process to easily escape without heating up the cave. If the door is very airtight it's worthwhile to also have a small vent low down on the door or wall to draw some fresh air into the cave.

Cold cupboards, another passive option

Another more permanent option (but for inside the house) is a "cold cupboard" or "cool cupboard". These are cupboards with a vent at the base that brings cold air from under the house up into the cupboard, with a small chimney at the top, usually painted black to draw heat upwards in order to keep the cupboard cool. David Holmgren's book *Retrosuburbia* has some details on this. Searching online for "cool cupboard" will find other useful information.

A biodynamic approach to cheesemaking

The biodynamic calendar is divided into four types of days, depending on the lunar zodiac sign. Fruit days are when the moon is in the fire constellations of Aries, Sagittarius or Leo. These are the best days for making cheese and other fermented foods. Flower days, when the moon is in the air constellations of Gemini, Libra or Aquarius also have good results, but fruit days are preferable.

Following the biodynamic calendar when you're making and aging cheese brings the benefit of being able to more easily plan your time. If you have a lot of tasks to do around the home, it is often overwhelming to face them all at once and not know which one to start with. The biodynamic calendar helps because we can look at it and say "this day is a fruit day, that's when I'll make this cheese, or tend to the cheeses", and structure other tasks around other days. The fruit days occur roughly every week, which is how often you should be tending to your aging cheeses. There's no need to be dogmatic if the right day is missed sometimes and you end up delaying the cheese tasks by a day or two, but in general it can be a good calendar to follow. Fruit days are also good for making fermented vegetables and preserving fruits - good things to keep you occupied in the kitchen while the cheese is being stirred.

The best time to make cheese

The best time to make hard cheese is naturally when you have lots of milk. If you have a limited amount of goats milk, you can arrange to have a cheese making day by not making any yoghurt, soft cheese or other dairy foods for a day or two, and limiting the amount of milk that's used for drinking. On cheesemaking days I generally set aside the bare minimum of milk we'll drink, and then make cheese with the rest. The length of time you can keep your milk for cheesemaking will depend on how cold you can keep it, how clean everything is, and how comfortable you feel with using older milk. Around three days was what I would use when I had a fridge. These days I don't have a fridge, and in summer raw milk will begin to get thick after around 24 hours, and to separate on its own into farmers cheese after around 36 to 48 hours, so I can get away with using milk from three milkings (morning, night, and next morning), if I start making the cheese early in the day.

Unless you don't have many responsibilities beyond cheesemaking, it's best to plan to be doing indoors chores or pottering around the kitchen when the cheese needs to be stirred. I like to plan to do my non-kitchen chores while the milk is culturing and setting into curds, and then to be doing a big wash up and kitchen clean while the curds need to be stirred. Many cheesemaking books will tell you to stir the cheese non-stop for forty five minutes, but this is something I can't imagine any home cheesemakers doing in practise. If you don't stir the cheese often enough, the curds will begin to knit together, but stirring around every five minutes is fine, and if they do begin to knit together it is not such a big deal, as you can just gently break them apart again and make sure to tend them more frequently next time. Alternatively, if you don't have many responsibilities or noisy toddlers around on cheesemaking day you could always read with one hand while stirring with the other, or listen to music or watch videos.

Cultured and renneted cheese recipes

Easy chévre

Makes around 1 1/2 cups of chévre

Ingredients
2 litres (2 quarts) goats milk
2 tablespoons milk kefir, viili, or fresh whey
1 tiny, tiny amount of rennet (see the method for details)
1 teaspoon unrefined salt without additives, or more to taste

Equipment
Jar or pot for culturing the cheese
Stirring spoon
Cheesecloth
Colander
Wooden spoon for hanging

Method
It's a lot easier to make a larger batch of chévre than it is a smaller one, but smaller ones naturally make a smaller amount of cheese that is easier to eat before it starts to go bad. Another advantage of a small batch is that it can be made when you only have a small amount of milk to spare. If you have a lot of milk, feel free to make a larger batch. I've kept the ingredients for this recipe to two litres, which is somewhere in the middle of these sizes. You can halve this recipe if you don't have much milk or many people to feed; you just need to be especially careful about the amount of rennet, as it's very easy to put too much rennet into a small batch of chévre.

Firstly get the milk at a good temperature for culturing. Milk straight from the udder strained into a jar is the best for this; you can just stir the culture and the rennet into the jar and leave it to culture until you see the curd separating from the whey.

If you're using cold milk, heat it up gently on the stove until it reaches a lukewarm temperature, around 72°F (23°C), then add the culture and rennet.

The rennet used in chévre is the smallest amount possible. Too much rennet will result in a rubbery curd. If you're using rennet tablets, scrape the tiniest tiny speck off a tablet and mix it into some water until it dissolves. If you think you may have used too much, throw away half this liquid or more. It's better to have too little rennet in this recipe than too much. If you're using liquid rennet, dilute a drop of it in some water, then use half of the diluted mixture in the cheese.

When the milk is warm, add the diluted rennet and the culture and gently but thoroughly mix it into the milk. Leave it to set at room temperature for around twenty four hours, until the curd has separated from the whey. You can leave it for a bit longer if you like, or less time if it has set quickly.

When the curd is ready, gently pour the curds and whey into a colander lined with cheesecloth. You can then tie the cheesecloth onto a wooden spoon and leave it to drain. Draining will take around eight hours. Once the cheese has drained, add some salt to taste, and either use right away as a spread, or drain for another couple of hours, to make a firmer cheese that stays fresh for longer.

Chévre will keep for a week or two in the fridge.

Fast mozzarella

There are two methods for making mozzarella. The first is to culture it slowly to develop the right amount of acidity to allow it to stretch. The second method is to add the acid as vinegar or lemon juice, at the right stage of the recipe. The first method will have more flavour, but is a sometimes unreliable process, the second makes a great cheese for pizza in around forty five minutes.

Making mozzarella the fast way means that if your family likes pizza and other melty cheese dishes, you can easily provide for this at home, without needing to devote hours to the process. It's possible to make fast mozzarella while the oven heats up and the dough rises.

Vinegar added to hot milk results in ricotta, so be sure to add the vinegar to your milk for mozzarella when cold or at room temperature.

This recipe makes around 350g (12oz) of cheese.

Ingredients
1 gallon (4 litres) milk
½ cup vinegar, diluted in 1 cup of water
¼ rennet tablet, or ¼ teaspoon liquid rennet, diluted in water
¼ cup unrefined salt without additives

Equipment
Pot with at least 4 litre (1 gallon) capacity
Colander and cheesecloth
Slotted spoon
A second spoon, either metal or wooden

Method
Gently stir the diluted vinegar into the milk before the heating begins. Heat the milk gently on the stove, while stirring often to a lukewarm temperature of around 90°F (32°C). Dissolve the rennet in around 1/4 cup of water, stir it constantly and then add it to the milk. Stir for two minutes, then leave the pot alone for 5 minutes. Check to see if a curd has formed. When the curd forms, cut it into one inch cubes.

Put the pot back on the heat and heat it to around 110°F (43°C) while stirring the curds. Remove it from the heat and continue to stir the curds for two to five minutes.

Remove the curds from the whey using a slotted spoon, drain.

Add 1/4 cup salt to the whey and heat it until it reaches a temperature of at least 175°F (80°C) but is not boiling.

Take handfuls of the drained curd, place on the slotted spoon, and lower into the hot whey for a few seconds, then use the slotted spoon with a second spoon to stretch and knead the curd several times, to develop strings, before forming into a ball shape. If it isn't stretchy at this point, try to put it in the hot whey for a bit longer before trying again.

Continue this for the rest of the curds, draining them as you go.

Mozzarella can be used right away, or will keep in the fridge for around a week.

Feta

Goats milk makes the best feta cheese. It can also be made in smaller batches than for the aged cheeses, so is a good cheese for backyard goatherds to make often.

Feta is traditionally aged in a salty whey brine; as long as the aging feta is kept below 10°C, the brine will preserve it for months. Feta can also be stored for a shorter time in the fridge, sprinkled with salt instead of brine.

Ripening time: 1 hour
Rennetting time: 1 hour
Stirring time: 30-60 minutes
First pressing: 30-60 minutes
Second pressing: 30-60 minutes

Ingredients
1 gallon (4 litres) milk
¼ cup milk kefir, viili or fresh whey
¼ rennet tablet, or ¼ teaspoon liquid rennet
Unrefined salt without additives

Equipment
Pot with at least 4 litre (1 gallon) capacity
Spoon
Knife for cutting curds
Pot or bowl, for storing the whey
Colander and cheesecloth
Cheese mould with follower and something heavy to weigh it down
1 litre jar with a glass or plastic lid, for brine-aging

Method
Gently heat your milk to a lukewarm temperature of around 90°F (32°C), thoroughly mix through the viili, then leave it to sit for an hour to ripen.

Dissolve the rennet in ¼ cup of water, stir it constantly and then pour it over the cultured milk. Quickly and thoroughly mix it through all the milk for a minute or two, then leave it to sit for another hour, until there is a clean break in the curd.

Once the curd has set, cut it into ¾" (2cm) cubes. Stir the curd every few minutes for the next 30 to 60 minutes, until they are firm and slightly springy.

Stop stirring and allow the curds to settle to the bottom of the pot, then pour the whey into a different pot or bowl; make sure to keep a litre of this for the brine.

Drain the rest of the curds through a colander, then mix through 2 tablespoons of unrefined salt. Leave the salted curds to drain for around ten minutes, stirring a couple of times to stop the curds sticking together too much.

Put cheesecloth in your cheese mould and put the salted curds in here; gently cover the cheese with the cloth, then the follower, and then around 5lb (2 1/2kg) of weight (a 2 litre jar filled with water works well), for around 2 hours, unwrapping the cheese, flipping it around, and re-wrapping every so often. You can get away with using a one litre jar for the weight if you fill it with warm water or warm whey.

While you're waiting for the cheese to press, you can make the brine by dissolving ¼ cup of unrefined salt without additives into a litre of warm whey. Set this aside until later.

Cut the cheese into pieces, either slabs or large cubes, whichever will fit in your brining container (don't cut them into little cubes like feta in jars in the shops, or you're likely to lose a lot of your cheese) then sprinkle salt all over the pieces, and leave them to air-dry for at least a day, until the edges seem to have developed a bit of a crust. Turn them around halfway through this time, to make sure that all the surfaces have a chance to air-dry.

Put the feta pieces in the brine and store below 10°C (50°F). It will be best if you can wait at least a week before eating it, longer is even better. You can get away with storing it at a warmer temperature if you eat it earlier and don't store it for long.

Halloumi

Halloumi is another cheese that can be aged in brine. It can then be fried to form a tasty sandwich filling or snack.

Halloumi can be made either with or without an added culture such as viili or yoghurt. The cultures already in raw milk are usually enough, but I like to throw a bit of kefir in too.

Rennetting time: 40-60 minutes
Stirring time: 20 minutes
First pressing: 30 minutes
Second pressing: 30 minutes
Heating time: 30-40 minutes

Ingredients
1 gallon (four litres) milk
Optional 2 tablespoons milk kefir, viili, yoghurt, or fresh whey
¼ tablet rennet, or ¼ teaspoon liquid rennet

Equipment
Pot with at least 4 litre (1 gallon) capacity
Spoon
Knife for cutting curds
Cheesecloth
Bowl or pot for catching whey
Cheese mould, follower, and weight

Method
Gently heat your milk to a lukewarm temperature of around 90°F (32°C). Dilute the rennet in 1/4 cup of water, stir it constantly, then add this to the milk, along with the optional viili, and thoroughly mix it through. Leave to set for around 40 minutes to an hour, until the curd cleanly breaks.

Cut the curd into one inch cubes.

Increase the temperature to around 38°C (100°F) while stirring often to make sure the curds don't stick to the pot. Keep the cheese at this temperature for another 20 minutes, while stirring every so often.

Allow the curds to settle for 5 minutes, before pouring off the whey into another pot or bowl to use later in this recipe.

Put the curds into a cheesecloth, inside the cheese mould, top with the follower, and then with some weight to help it press, such as a two litre jar filled with water. Leave it to press for around an hour, flipping it over halfway through to remove the cheesecloth, rewrap, and press with the other side of the cheese down.

While the cheese is pressing, heat the whey up until almost-boiling (around 90°C/195°F). Once the whey is hot, add the cheese and leave it in the hot whey, making sure it doesn't boil, until the cheese rises to the surface. This will take around 30-40 minutes.

The cheese can now be cooled, and either sprinkled with salt and fried right away (or kept for a few days in the fridge), or stored in a 5% brine. It will store for a longer time if kept under 10°C (50°F).

To make a 5% brine, add 50 grams of salt to a litre of water. This works out to be around 3 tablespoons per quart.

Tomme and Asiago

Tomme and Asiago are hard cheeses which can be made without a cheese press. For making them at home the only difference between the two is that Tomme is ripened by the lower temperature mesophilic bacteria, and Asiago uses the cooked curd process, to bring the curds to a higher temperature, encouraging the higher temperature thermophilic bacteria.

Tomme is the simplest hard cheese to make.

The taste of these cheeses is wonderful. Tomme and Asiago are not as familiar to us as some other hard cheeses are, so we don't have much of an expectation of what they should taste like, and can appreciate them more for what they are rather than trying to compare them with other cheeses. They have some similarities with alpine cheeses, with a complex, pleasant and satisfying taste that lends itself well to being sliced thinly and appreciated on its own, or used anywhere you would use a hard cheese. Asiago, with it's cooked curd method, tends to have more of the slightly sweet, nutty taste that comes from the slight caramelisation of the lactose during the cooked curd process (although this is more pronounced in a cows milk version of the cheese rather than ones made with goats milk). Tomme is a bit simpler, but still full of flavour, and is one of my favourite hard cheeses to make.

Ripening time: 1 hour
Rennetting time: 1 hour
Stirring time: 35-60 minutes
First pressing: 30 minutes
Second pressing: 12-24 hours

Ingredients
6 litres (6 quarts) milk
1/3 cup milk kefir, viili, or fresh whey
1/2 tablet rennet, or 1/2 teaspoon liquid rennet
Additive-free salt

Equipment
Pot with at least 6 litre (6 quart) capacity
Spoon
Knife for cutting curds
Cheesecloth
Bowl or pot for catching whey
800g (see page 113) cheese mould, follower, and weight

Method

Gently heat your milk to a lukewarm temperature of around 90°F (32°C), thoroughly mix through the viili, then leave it to sit for an hour to ripen.

Dissolve the rennet in half a cup of water, stir it constantly and then pour it over the cultured milk. Quickly and thoroughly mix it through all the milk for a minute or two, then leave it to sit for another hour, until there is a clean break in the curd.

Cut the curd into small cubes, around 1/5th of an inch (half a centimetre) in size. Allow the curds to rest for five minutes.

Slowly heat the curds to 38°C (100°F) over the course of half an hour, stirring often to make sure the curds don't clump together or stick to the bottom of the pot.

If making Tomme:

Maintain the temperature of 38°C (100°F), while stirring every so often. After around 5 minutes, begin testing the feel of the curds - they should feel slightly springy when squeezed. Continue to stir every few minutes while maintaining the temperature until the curds have reached the right texture. It may take up to half an hour.

If making Asiago:

These curds are heated to a higher temperature, so it's important to be more careful about stirring often than it is with Tomme. Over the course of half an hour, continue heating the curds while stirring often, until the temperature reaches 46°C (115°F). The curds should feel strong and have a spring to them when squeezed once they're ready.

To finish either cheese:

Once the curds have reached the right texture, allow them to settle to the bottom of the pot, and then pour off as much whey from the top as you can.

Line the cheese mould with cheesecloth and put the curds into this.

Fold the cheesecloth over the top, cover with the follower, then cover with a one litre jar filled with water or warm whey.

Allow to press for half an hour, then remove the cheese from the cloth, flip it over, and press again. If you can add more weight to this next pressing it will be better for it, if not, it will not matter so much.

Allow the cheese to press this second time for 12 to 24 hours.

Carefully remove the cheese from the cloth, rub 1 1/2 tablespoons of salt all over it, and allow it to dry in a cool, airy place for a few days, to begin developing a rind.

Once the rind has formed, move the cheese to wherever you're aging it. Age for at least two months, rubbing the rind with some salt or a clean dry cloth around once a week.

Havarti and Gouda

These cheeses use the washed curd method, which limits the forming of acids during the cheesemaking process, giving the cheese a mild taste.

Havarti doesn't need a cheese press, and the taste is similar to gouda, but the texture isn't quite as firm. While Gouda is usually covered in wax before aging, and Havarti is sometimes waxed as well, both these cheeses can easily be made with a natural rind.

To make measuring rennet simpler I've made these recipes using 8 litres of milk. If you don't have much milk you can scale the recipe down, using 1/3 of a rennet tablet for 6 litres of milk.

Be very careful about the source of water used in this recipe, to make sure you're not adding in any chemicals or bad bacteria. I use creek water or rain water that has been boiled and then cooled.

Ripening time: 30 minutes
Rennetting time: 30 minutes
Stirring time: 40 minutes
First pressing: 20 minutes
Second pressing: 3 hours

Ingredients
8 litres (2 gallons) milk
1/2 cup milk kefir, viili or fresh whey
1/2 tablet rennet, or 1/2 teaspoon liquid rennet
Additive-free salt
Optional 1/2 teaspoon caraway seeds, boiled for a few minutes

Equipment
Pot with at least 8 litre (2 gallon) capacity
Spoon
Knife for cutting curds
Cheesecloth
Bowl or pot for catching whey
800g (see page 113) or 1kg cheese mould, follower, and weight

Method

Gently heat the milk to 30°C (86°F). Mix the viili and the optional water from boiling the caraway seeds into the milk and leave to ripen for half an hour. Leave the caraway seeds for later in the recipe.

Dissolve the rennet in half a cup of water, stir it constantly and then pour it over the cultured milk. Quickly and thoroughly mix it through all the milk for a minute or two, then leave it to sit for another half an hour, until there is a clean break in the curd.

Cut the curds into 1.3cm (1/2") cubes. Allow the curds to rest for ten minutes, then stir them for five minutes.

Remove whey until the total volume in the pot has been reduced by around one quarter. Measure the amount that was taken out, and then add back in the same amount of hot water, at around 46°C (115°F).

Heat the pot until the temperature is 37°C (98°F), stirring the curds often as it warms. Continue to stir the curds every so often for another 40 minutes.

Check to see if the curds are ready by squeezing a handful of them; they should stick together.

Once the curds are ready, allow them to settle to the bottom of the whey for five minutes before pouring the whey off, mixing through the optional caraway seeds, and then putting the curds into the cheesecloth inside the cheese mould.

For making Havarti:

Wrap the curds in the cheesecloth, put the follower on top of the cheese mould, then cover with around 1kg (2lb) weight. Press for 20 minutes, then unwrap the cheese, flip, and rewrap, and then press with around 2kg (5lb) of weight for 3 hours.

For making Gouda:

You will need a cheese press. Wrap the curds in cheesecloth, put the follower on top of the cheese mould, then press at around 9kg (20lb) of weight for 20 minutes. Unwrap the cheese, rewrap, then press at around 22kg (50 pounds) of weight for 3 hours.

To finish both cheeses:

Rub 2 tablespoons unrefined salt over the surface of the cheese, and leave it to dry in a cool, airy place for a few days, to begin developing a rind.

Once the rind has formed, move the cheese to wherever you're aging it. You can begin eating this cheese after a month, but it will develop more flavour as it ages for longer.

As the cheese ages, check on it once a week, rubbing the rind with some salt or a clean dry cloth around once a week.

Alpine cheeses

Tomme and Asiago (page 128) are from places close to the Alps, but what generally is considered an alpine cheese is a cooked curd cheese that forms a rind for a few days at the usual cool temperature, and then is kept at a warmer temperature for around a week, to encourage holes called 'eyes' to form in the cheese, before being aged at the usual cheese cave temperature. It can be difficult to get eyes to form in smaller cheeses, and part of the flavour of traditional alpine cheeses comes from them being very large wheels that are aged for a long time, so I am usually content to enjoy Tomme and Asiago instead of alpine cheeses, but for anyone who wants to try a small batch alpine cheese, instructions are provided here.

Gruyére and Emmental are made in pretty much the same way. The rennet is added at around the same time as the culture, the curds form, then are cut and cooked. Gruyére goes through four pressings, being pressed with probably the highest weight that your cheese press will go to in the end. Emmental is pressed three times, making it a bit easier to organise being there to flip it at the right times. Gruyére is aged at the normal cheese cave temperature, where as Emmental goes through a higher temperature stage of aging, to encourage the hole-forming bacteria.

Jarlsberg can be considered an alpine cheese in some ways, but the way to make it is to follow the Gouda recipe on page 131, with the optional addition of a small piece of Jarlsberg or other alpine cheese, and then to follow the aging process for the Emmental option in this recipe, to encourage eyes to form in the cheese. Once the eyes have formed in the cheese, Jarlsberg is then usually eaten a bit earlier than Emmental, at around 2 to 3 months.

Rennetting time: 50 minutes
Stirring time: 65 minutes
First pressing: 30 minutes
Second pressing: 30 minutes-2 hours
Third pressing: 12 hours

Ingredients
8 litres (2 gallons) milk
1/2 cup milk kefir, viili, fresh whey, or yoghurt
Optional small piece of Gruyére or Emmental
1 rennet tablet, or 1 teaspoon liquid rennet
Additive-free salt

Equipment
Pot with at least 8 litre (2 gallon) capacity
Spoon
Knife for cutting curds
Cheesecloth
800g (see page 113) or 1kg cheese mould with follower
Cheese press

Method
Gently heat your milk to a lukewarm temperature of around 32°C
(90°F). Thoroughly mix through the viili, and leave it to sit for a
minute or two. At this time you can also add part of a good alpine
cheese, to bring some of the hole-making culture into your cheese,
but this is optional. To do this, use a very fine cheesegrater to gently
sprinkle your milk with cheese 'dust', or you can soak the piece of
cheese overnight in a small amount of milk, and then remove the
cheese, adding the cheese-cultured milk to the recipe with the viili.
If you want to do this, make sure that the cheese you get is very fresh
and doesn't have preservatives added.

Dissolve the rennet in half a cup of water, stir it constantly and then
pour it over the milk. Quickly and thoroughly mix it through all the
milk for a minute or two, then leave it to sit for around fifty minutes,
until there is a clean break in the curd.

Cut the curds into 5mm (1/4") cubes.

Slowly heat the curds to 49°C (120°F) over 35 minutes while stirring
often. The curds should shrink a bit.

Keep the temperature at 49°C (120°F) for another half an hour,
stirring every few minutes. When the curds are ready they will bind
together when you squeeze a handful of them.

Allow the curds to settle, and pour off most of the whey. Line your
cheese mould with cheesecloth and put the rest of the curds in
here. Fold the cheesecloth over the top, cover with the follower, and
press with around 4 1/2kg (10 pounds) of weight for half an hour.

To make Gruyére:
Remove the cheese from the mould and cloth, flip it over, rewrap and then press from the other side with 7kg (15lb) of weight for another half an hour.

Remove the cheese, unwrap, flip and rewrap, press at 18kg (40lb) of weight for 6 hours.

Remove the cheese, unwrap, flip and rewrap, then press at 22kg (50lb) of weight for 12 hours.
Remove the cheese from the cloth, rub 2 tablespoons unrefined salt over the surface of the cheese, and leave it to dry in a cool, airy place for a few days, to begin developing a rind.

Once the rind has formed, move the cheese to wherever you're aging it. It will develop more flavour as it ages. It's best to wait at least 6 months if you can, but you can start eating after 2 months if you really want.

As the cheese ages, check on it once a week, rubbing the rind with some salt or a clean dry cloth around once a week.

To make Emmental:
Remove the cheese, unwrap, flip and rewrap, press at 7kg (15lb) weight for 2 hours.

Remove the cheese, unwrap, flip and rewrap, then press at 7kg (15lb) weight for 12 hours.

Remove the cheese from the cloth, rub 2 tablespoons unrefined salt over the surface of the cheese, and leave it to dry in a cool, airy place for a week, to begin developing a rind.

Once the rind has formed, move the cheese to a place with a temperature of around 20-23°C (68-74°F) with fairly high humidity (a closed container can be used to increase humidity) and age it there for two to five weeks, wiping the cheese with a brine-soaked cloth or some salt every day, and turning it over each day, making sure it's on a dry surface each time. The cheese should expand during this time, giving a rounded barrel-like look to the sides of the wheel.

Once the cheese has expanded (or five weeks is over), bring the cheese back to the usual aging space, preferably at around 7°C (45°F) and age for another four months or more.

As the cheese ages, check on it once a week, rubbing the rind with some salt or a clean dry cloth around once a week.

Bloomy white rind cheeses

Success with white bloomy rinded cheeses without packaged cultures will depend on what flora is present in the milk and the cheesemaking environment. I have tried to make bloomy rinded cheese in the past when my goats were eating mostly lucerne without success - I followed the recipe exactly, which told me that the raw milk and the natural culture I was using would already have enough fungus to culture the surface. I gave the cheese all the right conditions to mature, the white rind just did not grow, and the smell was horrible.

Now that my goats are eating a diet based on trees (mostly white maple leaves at the time of writing this) white rinded cheeses seem to make themselves. Some farmers cheese sitting in the switched-off fridge that I was waiting to use in a cheesecake grew a perfect coating on the top and had the familiar mushroom-like scent of Camembert. An attempt at an alpine cheese that was stuck to the cheesecloth and abandoned on the kitchen bench for a couple of days was unwrapped to reveal a pleasant surprise of camembert aroma and white fuzzy bits on it, the texture was a bit strange to eat it as is, but when it was melted it reminded me of Brie melting on top of toast. Natural cheesemaking brings all kinds of surprises.

To encourage white rinds to grow you can bring some Camembert or Brie in from the shop and put it next to the cheese that you want the white rind to grow on. You can also gently blow at the surface on the Camembert in the direction of your cheese, to help spread the spores, or add some of the rind to your milk during the culturing process, removing it before the rennet is added. It may be worth experimenting with aging some chévre or another small batch of cheese without bringing in any Camembert from the shop, just to see if your milk and culture will already grow the right fungus without any effort.

Camembert and Brie

The original versions of these cheeses have many differences, mainly from the difference in what the cows graze on, as well as the size of the cheese, with Brie being larger and therefore taking longer to age and developing different flavours to Camembert. Outside of France, any locally made cheese labelled as Brie won't be made any differently to the same cheese called Camembert, especially if they're both made in small wheels. If you already make hard cheeses you will have a mould that can be used to make a Camembert, or maybe one that will make a cheese in between the sizes of Camembert (10cm or 4") or Brie (18cm or 7"). If you don't have one, you can have a look around the house to see what can be used to make either smaller wheels of Camembert, or a larger wheel of Brie. Clean wooden baskets might do the job, as might the basket of a salad spinner. Yoghurt buckets with straight sides can have lots of holes punched in them, plastic yoghurt jars with sides that go in at the top can be trimmed and then have holes put in them. Cylinder moulds with open bottoms will make the draining process easier, but with careful handling basket-type moulds can be used as well.

If you don't have much milk, or if you want to try a smaller batch, you can easily halve this recipe and make one Camembert instead of two or three. You can also double this recipe if you want to make a bigger Brie.

Ripening time: 1 hour
Rennetting time: 1 1/2 to 3 hours
First settling: 12 hours
Second settling: 12-24 hours

Ingredients
4 litres (1 gallon) milk
1/4 cup milk kefir, viili, or fresh whey
Optional small piece of Camembert or Brie
1/4 rennet tablet, or 1/4 teaspoon liquid rennet
Additive-free salt

Equipment
Pot with at least 4 litre (1 gallon) capacity
Spoon
Knife for cutting curds
Either 2 or 3 10-12cm (4-5") cheese moulds for Camembert,
or 1 18cm (7") mould for Brie

Method

Gently warm the milk to 32°C (90°F). Stir the viili through the milk, add the small piece of Camembert rind now if you are using it. Leave to culture for an hour before removing the optional piece of cheese.

Dissolve the rennet in 1/4 cup water, stir it constantly and then pour it over the cultured milk. Quickly and thoroughly mix it through all the milk for a minute or two, then leave it to sit for an hour and a half, or as long as three hours. The curd will have reached the clean break point earlier than this, but we leave it for longer to develop the curd without stirring it.

Cut the curd into 5cm (2") cubes if making Brie, 2cm (1") curds if making Camembert, then ladle directly into the moulds. Keep some of the whey for making a brine to wipe the cheeses with.

Make the brine by combining a teaspoon of salt with 300ml (10oz) of whey.

Leave the cheese to drain in the mould for 12 hours. It should be reduced in volume by half. Carefully flip the cheese over, returning it to the mould for another 12 or 24 hours, or until the cheese is firm enough to keep its shape. The draining process will be most successful in a fairly warm temperature (around 21°C or 70°F).

Rub the surface of the cheese with salt, around 1 tablespoon in total, then leave in a cool, airy place, until the surface appears dry, this should take around 24 hours. Flip the cheese over a couple of times during this process, moving it to a dry surface each time.

Once the cheese is dry, age it in a high humidity space at the normal cheese cave temperature. For the first week of aging, wipe the surface of the cheese every second day with the salted whey brine, this will help to increase the humidity to encourage the white fungus to grow, while discouraging other blooms on the rind.

Once the first week has passed, stop washing the rinds and carefully flip your cheeses over twice a week, transferring to a dry surface each time. Continue doing this for three weeks. If any fungus other than the white one is growing, you may need to salt the surface of the cheese.

After a month of cheese cave aging has passed, there should be a coat of white fungus all over the cheese. You can now wrap your cheeses in beeswax wrap or baking paper (or you can use edible leaves such as grape leaves), and continue to age the cheeses at a lower temperature such as in a fridge. Camembert will be ready after a month in the fridge, Brie in two months.

Aged Chévre

Any cheese can be encouraged to form a bloomy white rind. Hard cheeses will end up with a subtle white dust over them, where as soft chévre ends up with one with more similarities to Camembert. To make an aged chévre you will need to drain your chévre until it can easily be shaped, but won't crumble to bits.

To make aged chévre, follow the chévre recipe on page 120 until it's time to drain the curd. You can then either continue to drain the curd in an 800g hard cheese mould, to make a Camembert-sized chévre, or you can find a couple of smaller soft cheese moulds and drain it in them. Alternatively, you can drain the curds in the cheesecloth as if you were making a normal chévre, and then shape the cheese into a log for aging.

Allow the chévre in moulds to drain for 24 hours, carefully flipping them over roughly halfway through, then remove them from the moulds, and salt with around 1 1/2 teaspoons of salt for every 2 litres of milk used in the batch.

Move the salted cheeses to a cool, airy place and allow them to air dry for around 24 hours, flipping once or twice during this time and moving the cheese to a dry surface each time.

Move the cheeses to a humid place (e.g. a closed container) at the regular cheese aging temperature, next to a piece of bloomy rinded cheese (blow dust from a white-rinded cheese over them if you can). Carefully flip them over once a day, moving them to a dry surface each time.

After around a week to a week and a half in the humid place, you can now move them to the regular cheese cave conditions, flipping them over every second day. Observe the cheese all through the aging process, if it begins to grow any fungus that isn't white, feel free to wipe it with a salty whey brine, or gently rub salt on it.

Aged chévre can be eaten after only a week in the cheese cave, or aged for up to two months.

Blue cheeses

Any cheese can become a blue cheese by introducing it to the penicillium roqueforti mould from other blue cheeses and offering it the right conditions to grow. At first a blue rind is encouraged to form, and then holes are poked through the cheese, to get the blue fungus on the rind to colonise veins in the cheese. In this way, you can take any cheese recipe and turn it into blue cheese.

Blue cheeses are usually not pressed, in order to have more gaps in the cheese for the blue fungus to grow. The curds can be salted and partly drained before being put in the mould; this dries out the curd, making for a cheese that can be aged for a longer time and develop more flavour.

To extract the blue fungus from a piece of cheese to culture your new cheese with, start with a wedge of cheese that has a good amount of blue on it. Choose a cheese that you know hasn't been treated with preservatives, and for best results, find one that has been freshly cut. Carefully slice off an outside slice of this, to expose a fresh side of cheese that no person or wrapper has ever touched. Use a spoon or knife to dig out a piece of this fresh side of cheese, choosing a piece that has plenty of blue mould on it, around the size of an almond is fine. Take the piece of cheese and steep it in some water, breaking up the blue bits as much as you can to get them mixed into the water. Strain this water into the milk when you're adding the viili culture, then follow the rest of the recipe.

Gorgonzola and Stilton

Ripening time: 30 minutes
Rennetting time: 1 hour
Resting time: 10-30 minutes
First draining: 10-15 minutes
Second draining: 12-24 hours
(Stilton) Third draining: 2 hours (flipped every 20 minutes)
(Stilton) Fourth draining: 4 days (flipped twice a day)

Ingredients
A small piece of blue cheese
8 litres (2 gallons) milk
1/2 cup milk kefir, viili, or fresh whey
1/2 a rennet tablet, or 1/2 teaspoon liquid rennet
2 tablespoons additive-free salt

Equipment
Pot with at least 8 litre (2 gallon) capacity
Spoon
Knife for cutting curds
Cheesecloth and colander
Spoon, for hanging
(Stilton) 2 chopping boards or baking sheets
(Stilton) 4 1/2kg (10 pounds) weight
Cheese mould, preferably open-ended on both sides

Method
Follow the directions in the third paragraph under the 'Blue Cheese' heading on page 143 to extract the blue fungus from a piece of blue cheese.

Gently heat the milk to around 31°C (88°F). Stir through the strained blue cheese water and the viili. Leave it to culture for half an hour.

Dissolve the rennet in 1/4 cup water, stir it constantly and then pour it over the cultured milk. Quickly and thoroughly mix it through all the milk for a minute or two, then leave it to sit for around an hour, until the curd cleanly breaks.

For Gorgonzola
Cut the curds into 1cm (1/2") cubes. Leave them to rest for ten minutes, then put them in a cheesecloth-lined colander.

Mix through 2 tablespoons of salt into the curds, then leave them to drain for 5 minutes, stirring often to stop them from knitting together.

Put the curds into a cheese mould and allow to drain for 10 minutes. Gently flip the cheese over, and then leave it to drain for 24 hours, flipping it over every so often to encourage it to drain evenly, and for the top and bottom to be flat.

Remove the cheese from the mould and move it to the usual cheese aging space.

For Stilton

Cut the curd into around 1cm (1/2") cubes. Allow the cut curds to rest for half an hour.

Pour the curds into a cheesecloth-lined colander. Once most of the whey has drained, tie the corners of the cheesecloth together and hang it from a spoon over a bowl or pot as you would for chévre, making sure the bottom of the cloth doesn't rest in the dripping whey. Allow the curds to drain in this way for 15 minutes.

Transfer the cheesecloth bundle to a chopping board or baking sheet without a rim set over a sink so that it can drain of whey as it presses. Cover the bundle with another chopping board or baking sheet, then place around 4 1/2kg (10 pounds) of weight on the top. Allow the curds to drain in this way for around 12 hours.

Break the curd into pieces around 2 1/2cm (1 inch) in size and transfer them to a bowl. Thoroughly mix through two tablespoons of salt.

Transfer the salted curds into a cheese mould (a cylinder-type mould will be much easier to work with than one with a base). Put a sushi mat or wooden board underneath it if you are using a cylinder-type mould. Flatten the top of the curds to make it even on top, then allow the curds to press under their own weight for 2 hours, carefully flipping it over every 20 minutes or so.

Leave the cheese in the mould for another four days at room temperature, flipping it over twice a day.

Remove the cheese from the mould and move it to the usual cheese aging space.

To finish both cheeses

After two or three weeks it should have developed a decent amount of mould. Use a sterilised teaspoon to gently scrape excess mould off the surface of the cheese, then use a sterilised metal skewer or metal knitting needle to pierce into the middle of the cheese from all around the sides of the cheese (but not too close to the top and bottom).

Return the cheese to the cheese cave and leave it to age for at least two months, flipping it over and moving it to a dry spot twice a week.

Traditional Cheddar

Cheddar cheese originated in a time and place where there were a lot of cows to be milked, and a lot of dairymaids around who needed something to do in between the morning and evening milkings. Real cheddaring is a long process that is ideally suited to more milk than you're likely to get from some backyard goats, but there is nothing stopping you from trying out a smaller batch of it, and I will provide instructions for this. Please feel free to double or triple the recipe if you have enough milk to do this.

Traditional Cheddar is covered in lard then wrapped in cloth, to stop the surface ripening bacteria from having an effect on the cheese, while still allowing the cheese to 'breathe' more than it would if it were covered in wax, making for a cheese that is ripened from the inside out to bring out different flavours to natural-rinded cheese. There is a noticeable difference between Cheddars that have been ripened in cloth and Cheddars from the same cheesery that are ripened in wax or plastic.

Ripening time: 40 minutes
Rennetting time: 30 minutes
Stirring time: 1 1/2 hours
Cheddaring time: 30 minutes
First pressing: 30 minutes
Second pressing: 30 minutes
Third pressing: 12 hours

Ingredients
8 litres (2 gallons) milk
1/2 cup milk kefir, viili, or fresh whey
1/2 a rennet tablet, or 1/2 teaspoon liquid rennet
2 tablespoons additive-free salt

Equipment
Pot with at least 8 litre (2 gallon) capacity
Spoon
Knife for cutting curds
Cheesecloth and colander
800g (see page 113) cheese mould with follower
Cheese press
Cheesecloth and lard for clothbinding the cheese

Method

Gently heat your milk to a lukewarm temperature of around 31°C (88°F). Thoroughly mix through the viili, and leave it to ripen for 40 minutes.

Dissolve the rennet in half a cup of water, stir it constantly and then pour it over the milk. Quickly and thoroughly mix it through all the milk for a minute or two, then leave it to sit for around thirty minutes, until there is a clean break in the curd.

Cut the curds into 1cm (1/2") cubes.

Slowly heat the curds to 39°C (102°F) over 45 minutes while stirring often. The curds should shrink to around the size of peas.

Keep the temperature at 39°C (102°F) for another 45 minutes, while stirring gently and frequently. When the curds are ready they will feel springy.

Allow the curds to settle, and pour off as much whey as possible. Drain the rest of the whey from the curds in a cheesecloth-lined colander. Return the curds to the pot, and keep the heat at 39°C (102°F) for the next half an hour while you carry out the cheddaring process.

You can either do a stirred curd Cheddar, by stirring the curds around every 5 minutes during this time, or you can make a traditional Cheddar by following the instructions below.

For traditional Cheddar:

Allow the curds to knit together for a few minutes, then cut it into slabs around 5cm (2") thick. Stack as many as will go on top of each other without toppling over. You can cut the slabs in half to make smaller pieces to work with if you want.

Move the bottom slab to the top constantly (or as often as you can), to allow all the slabs to evenly press under their own weight while the acidity increases. Make sure to keep the curds warm during this time, ideally between 32°C and 39°C (90°F and 102°F).

To finish either stirred curd or traditional Cheddar:

The curds will be ready when they have a strong rubbery texture, similar to a cooked chicken breast.

Break the curds up into pieces no bigger than 2.5cm (1"), add the salt, stir to combine, and leave to sit for 10 minutes while keeping the temperature warm to expel more whey.

Put the curds into a cheese mould lined with cheese cloth and press at half the maximum pressure for half an hour.

Remove the cheese, flip, and rewrap, then press for another half an hour at the same pressure.

Remove the cheese, flip, and rewrap, then press at the maximum amount of pressure for up to 12 hours.

Remove the cheese from the mould and allow it to dry in an airy place for a day or two, flipping it every so often, placing it on a dry place each time.

Bind your cheese in cloth by preparing four pieces of cheesecloth that will fit over the top and bottom of the cheese with an inch of excess cloth to go around the sides, also prepare two rectangles of cheesecloth to bind the side of the cheese.

Smear lard on the top of the cheese, press one piece of cheesecloth on this, smear more lard where the excess cloth needs to stick on the sides, and press the rest of the cheesecloth into this. Smear the cheesecloth with more lard and stick the second piece of cheesecloth over the top of this. Flip the cheese over and repeat for the other side, then smear the sides of the cheese with lard. Place one piece of cheesecloth around the sides, smear that with lard, and finish with the second rectangle of cloth.

Age the Cheddar for at least three months in a cheese cave. It will develop more flavour as it ages. Some people prefer milder Cheddars, aged for around 3 months, others prefer 6 months of aging. After 12 months Cheddar has a sharper flavour which some people prefer, and it can be aged even longer than this.

Keep an eye out for mould on the cloth during the aging process. Wipe off any mould that forms off with a cloth dipped in vinegar.

Other dairy products

Kefir

To make real kefir, I recommend tracking down some milk kefir grains. These can be gained for free from people who make milk kefir themselves, or purchased online through shops that sell cultures. There is another form of kefir called water kefir, the grains for these have a different appearance to milk kefir, with milk kefir grains looking like tiny cauliflowers, and water kefir grains like translucent squares. Make sure you get the milk kefir grains to make this recipe.

To make milk kefir, add a teaspoon of freshly rinsed milk kefir grains to one cup of milk, leave it to ferment at room temperature for eight to twenty four hours. The fermenting time will depend on how active your grains are, the ratio of grains to milk, and the temperature.

After the kefir has fermented as much as you want it to, remove the kefir grains and rinse them with cold water, and then begin your new batch, or put them in a little bit of milk in the fridge to be stored for later. It's important to rinse kefir grains in this way in between batches, to avoid any off-flavours developing from overfermented milk that might have stuck to the kefir grains.

The taste of kefir is not for everyone. If you don't like the taste to begin with, try fermenting for a shorter amount of time or reducing (or increasing) the amount of grains used, using milk from a different goat, or adding a teaspoon of honey and a pinch of cinnamon to every cup of kefir for drinking.

You don't have to enjoy the taste of kefir to make cheese with it. Just make sure you have some fresh kefir without anything added to it for your cheesemaking if you want to use it as a starter. You'll only need around half a cup of kefir for a hard cheese made from eight litres (two gallons) of milk, even less if you're making a small batch of soft cheese. You can store the kefir grains in the fridge in a small amount of milk; just rinse them and then begin a new batch of kefir 8-12 hours before you want to make cheese. If the kefir has been in the fridge for quite a while, it may need to be rinsed and fed a second time before it's ready to use.

Yoghurt

The thickness of goat yoghurt will vary depending on which goat it's made from. My Toggenburgs always made fairly thin yoghurt, the Saanen/Nubian cross that I look after makes the best thick yoghurt, and the pure Saanen yoghurt is somewhere in the middle of these. If you mix all your milk together you may get decent yoghurt, or you may not, so it can be worth trying each goat's milk one by one to see which is the best. Winter milk also makes thicker yoghurt than summer milk. Feeding barley to the goats will increase the thickness of their milk.

The easiest kind of yoghurt to make from fresh raw goats milk is viili, which is a Finnish culture that is added to milk warm from the udder, and left at room temperature for 24 to 48 hours until it's thickened and soured.

Viili should be made at least once per week (preferably twice), to keep the culture fresh. It is worth keeping a small amount of viili tucked away in the freezer, just in case your yoghurt gets contaminated or abandoned.

To make viili, start with a jar around ¾ full of milk, then add a plonk of viili from a previous batch, put the lid on, and gently shake it a bit to mix it through. The right ratio is around 5 tablespoons of viili to one litre of milk but there's no need to be exact about it, just don't put too little in or the viili culture may not be able to multiply enough before the cultures already in the milk take over. If you only have a tiny amount of viili to begin with, just make a smaller batch and then make a bigger one once the small batch is finished.

If the yoghurt is not thick enough for your taste, you can ferment it for longer, until the whey separates, and then drain it through some cheesecloth as you would for chévre until it's as thick as you'd like it to be.

In winter I put the yoghurt on a warm (not hot) part of the wood stove and let the yoghurt gently warm up a bit (until the bottom of the jar feels nicely warm when put on the inside of the wrist), and then move it to a fairly warm part of the house to finish fermenting. Sometimes it needs a second or third go on the wood stove. In warmer weather (or warmer houses), normal room temperature is fine. 24 hours is usually all the time that's needed to culture the viili in warm weather, 48 hours is more realistic when it's cold.

You will need to track down some viili culture to make this yoghurt. I found mine on eBay. Cultures for Health also sell viili culture.

You can try making yoghurt with regular shop-bought yoghurt with live cultures. To make this you first need to heat the milk until boiling point and then let it cool to around 40°C (104°F). Stir through the yoghurt (around the same ratio as you would for viili). You need to find a way of keeping this at 40°C (104°F) for the next 6 hours. Some ways of doing this are: fermenting in an insulated food jar such as a thermos, or in the warming oven of an old woodstove, or surrounded by hot water or hot water bottles in an esky. Fresh yoghurt as a starter will have a better result than old yoghurt.

Custard

You don't need cream to make custard, but if you can use milk from the goat that gives the creamiest milk then it will be better.

Ingredients
2 cups milk (the creamier, the better)
1 or 2 tablespoons tapioca flour (aka arrowroot starch)
1 or 2 tablespoons coconut sugar or honey
A pinch of vanilla bean powder, or a teaspoon of vanilla extract
5 egg yolks

Method
Heat the milk up, stirring every so often, until it feels very hot and is about to boil.

While the milk is heating up, in a measuring jug or bowl, mix together the tapioca flour, coconut sugar, vanilla bean powder and egg yolks until evenly combined.

Pour a little of the hot milk into the egg yolk mixture and mix until evenly combined, then add a bit more, mixing through again, and then add as much as will fit in the jug. Pour the mixture back into the saucepan, mixing thoroughly if there's any more milk left in there. Return to a low-medium heat and continue to heat up, while stirring, until thickened and hot. Serve hot, either on its own or as a side to steamed puddings, stewed fruit and crumbles, or serve it as a cold dessert.

Meal recipes

Chévre (or quark) pastry

A simple pastry that can be fermented overnight to get more nutrients from wholegrain flour, or you can use unbleached white flour instead if you prefer. The pastry can be rolled out and used for savoury tarts and pies or for sweet pastries.

Ingredients
250g (9oz) butter, at room temperature
250g (9oz) chévre (page 120) or quark (page 106)
300g (10.5oz) wholegrain spelt or wheat flour (or white spelt flour)

Method
Cream the butter in a large bowl, then stir through the chévre, a little at a time, to make sure that it combines evenly.

Gently stir though the flour, stirring as little as possible. Then cover the bowl with a damp tea towel and allow the dough to rest at room temperature overnight if using wholegrain flour. Once the wholegrain dough has rested (or the white flour dough has been mixed), put it in a cold place such as a fridge and leave it to chill for at least half an hour before rolling it out to bake.

If you want the pastry to be as flaky as possible, roll it out on a floured surface, then fold it in thirds like a letter, then roll it out again, and fold it again. It won't be as flaky as puff pastry, but it is a lot tastier.

Palak paneer

Paneer is fried and served in a mildly spiced spinach sauce for a quick and nourishing meal. Serve on its own, or with flatbread or rice. When I want to make a one pot meal I sometimes add some sliced turnips to the sauce before it simmers, and they cook with the sauce. Cooked potatoes can also be added. Any kind of green can be used, just cook it for as long as you would usually cook the green. Baby spinach is ready in a few seconds, kale and turnip greens take longer.

This recipe serves two and can easily be doubled or tripled.

Ingredients
2 tablespoons lard or ghee
1 teaspoon brown mustard seeds
1 small onion, finely chopped
1 clove garlic, finely chopped
1cm (1/2") piece ginger, finely chopped
Optional 4 curry leaves
1/2 teaspoon coriander (ground)
1/2 teaspoon cumin (ground)
1/2 teaspoon garam masala
1/4 teaspoon turmeric
Optional pinch of asafoetida
1/2 cup tomato passata
1/2 cup milk
1/2 teaspoon honey
1/2 teaspoon salt
Paneer (page 102), or a double batch if you're super hungry, cut into cubes
200g (7oz) spinach or other greens, finely chopped

Method

Heat the lard in a saucepan over medium heat and add the mustard seeds. After around 30 seconds, when they've started to pop, stir in the onion for a minute or two, before adding the garlic, ginger, curry leaves, coriander, cumin, garam masala, turmeric and asafoetida.

Cook for around 2 minutes, until the onions are beginning to smell cooked, and are breaking down.

Stir through the passata and milk, continue to cook over a medium-low heat for 10-15 minutes. Add the honey and salt, and taste to see if it needs any more salt, or a tiny splash of cider vinegar.

While the sauce is simmering, fry the paneer on a greased frypan or the hotplate of a combustion stove over medium-high heat. Cook until browned on one side, then flip over and cook on the other side.

Once the sauce has simmered and the paneer is ready, quickly stir the spinach through the sauce. If it's spinach then it will be ready in around 30 seconds, if you're using a different green it will need to cook for longer. Once the greens are cooked, gently stir the cooked paneer through, and then serve.

Chévre and spinach crustless pie

This is a great recipe to use up extra chévre that's getting a bit old to eat raw. More of an idea rather than a recipe, you can adjust the ingredients depending on what you have on hand, and increase or decrease the amount of eggs depending on how many people you want to feed.

If you use lots of chévre it will be like a spinach and ricotta pie, less chévre will make it more like a frittata. Made either way, it tastes great and is a wonderful way to make a meal from easily homegrown foods.

Serves two or four, depending on the amount of eggs used.

Ingredients
1-2 tablespoons lard or butter
1 small or medium onion, finely chopped (or use leeks or spring onions)
A few leaves of silverbeet (chard), spinach, or other green
A small handful of fresh parsley, optional
A couple of leaves of sage, optional
Salt and pepper to taste
4 to 10 eggs
Leftover chévre, as much as you have
Optional leftover roasted or boiled potatoes (or other roast vegetables), thinly sliced

Method

Heat the lard in a skillet on the stove over medium heat. Preheat the oven to a reasonably hot temperature (around 180°C/350F). When the lard has warmed up, add the onions, mixing them thoroughly through the lard, and cook them, stirring every so often, until they smell good, and are either turning golden, or beginning to break down.

While the onions are cooking, chop up the herbs and greens. Keep the stems separate from the leaves, then once the onions are cooked, stir through the stems for a few minutes, until they are almost cooked, then stir through the leaves for around a minute or two, until they have lost a lot of their volume. If there is excess water in the pan, cook it for longer to evaporate the water, or tip the water out. Mix through a pinch of two of salt and taste some of the vegetables, adding more salt if they don't taste salty enough to eat on their own. Remove the pan from the heat and allow the vegetables to cool slightly while you prepare the eggs.

Whisk the eggs together in a mixing bowl until no traces of yolk or white remain. If you have lots of chévre, you can mix some of it in now, keeping the rest for dotting on top of the pie. Add a couple of grinds of pepper.

Add the cooked vegetables to the egg mixture, scraping the pan clean, then return the pan to the heat with a little more lard brushed over it to stop the pie sticking.

Pour the egg mixture into the hot pan, place the slices of cooked potatoes evenly over the mixture, then dot with chévre. Put in the oven and cook until set and lightly browned on the top.

Serve with salad, fermented vegetables, or homemade chutney or relish

Paneer stir fry

More of an idea than a recipe. You can make a sweet and sour one by mixing the juice of an orange with a tablespoon of honey and some coconut aminos or tamari, to taste, or you can make a plain stir fry just by drizzling it with coconut aminos and maybe a little honey and salt, to taste.

This recipe can be adjusted to feed as many or as few people as you want. Adjust the size of your pan depending on how many people you are feeding. The vegetable amounts are given for a recipe that will serve 2 to 4 people.

Ingredients
3 tablespoons lard
Paneer, cut into cubes, as much as you want to use
1 small or medium onion, cut into eight wedges
At least 1 carrot, julienned
Optional clove of garlic, finely chopped
Stir fry vegetables of your choice (see note at end of recipe)
Coconut aminos, tamari or other fermented sauce, to taste
1 tablespoon honey
Optional juice of one orange, to make it sweet and sour
Salt, to taste

Method
Cut paneer into cubes and brown them over medium-high heat in a little lard, flipping over halfway through cooking to cook the other side. Set aside until the end of the recipe.

Add extra lard if the pan looks dry, it should have a healthy layer on the bottom, but the vegetables shouldn't be swimming in it. Continue to heat the pan over a fairly high heat, and place the onion wedges, cut side down on the pan. Allow them to cook for a few minutes, until very deep brown on the cut side, then throw in the carrots and stir everything around.

Allow the onions and carrots to sit there cooking, stirring every now and then, until the carrots are almost done; they should be crispy but cooked. If you have used a julienne peeler to cut them then the greens can be added right away; if they are big chunky sticks of carrot they will need to be cooked for a few minutes.

Throw in the rest of the vegetables and cook while stirring, until they are almost cooked to your liking. The length of time this will take depends on the choice of vegetables, and how finely they are cut.

Add the sauces of your choice and stir through, adjusting the seasonings to taste, and adding some water if it seems too dry. Gently stir through the paneer, coating it in the hot sauce to warm it up. As soon as it seems warm enough, serve the stir fry right away, either on its own, or with some rice.

Recommended vegetables: I find this is a satisfying recipe just with onion, carrot and finely shredded cabbage, but it is also a good way to use up all kinds of vegetables such as broccoli, capsicum, green beans, snow peas, bok choy, and anything else that cooks quickly and needs to be used. The amount of vegetables to add is up to you, and will depend on whether you're serving it with rice or another side dish. I often just add lots of cabbage and carrot, and these act as 'noodles'. Chopped pears or pineapple can be added to the sweet and sour version.

Cheesecake

This makes a traditional German-style cheesecake. Nourishing, full of flavour and not too sweet. It can be served as-is, or topped with fruit or jam of your choice.

Ingredients
For the crust:
1 1/2 cups white spelt flour
1/2 teaspoon bicarb soda (baking soda)
A pinch of salt
A pinch of vanilla powder (optional)
1 tablespoon finely grated lemon zest
1 egg, whisked
1/4 cup coconut sugar
100g (7 tablespoons) butter

For the filling:
3 eggs, separated
3/4 cup coconut sugar
A pinch of vanilla powder, or a teaspoon of vanilla extract
100g (7 tablespoons) butter, soft
3/4 cup cream
500g (17oz) quark or ricotta
1 tablespoon tapioca flour
A pinch of salt

Method

Make the crust by mixing together the flour, bicarb, salt, vanilla, lemon zest and coconut sugar. Rub the butter in until the mixture resembles breadcrumbs, then mix through the egg and briefly knead to make a dough. Divide the dough into three pieces, press two of these into the bottom of a buttered 9" springform pan, pressing up into the sides slightly, and then take the remaining piece, roll it out in your hands into a few sausage shapes, and then press them into the sides of the tin, pressing as high up as you can go, while making sure the dough has all joined together. Put the cake tin somewhere cold for the pastry to firm up.

Get the oven to around 150°c (300°f).

Make the filling by beating the egg yolks with the coconut sugar until pale and foamy. Add the soft butter and continue to mix, and then add the cream and the quark; mix until evenly combined.

In a separate bowl, whisk the egg whites with the salt until stiff, then gently fold these into the quark mixture while adding the tapioca flour as a sprinkle.

Bake in the oven for 50-60 minutes, until golden and set, then switch the oven off and leave the cake in there with the door shut for another 15 minutes.

Remove the cake from the oven and allow it to cool down at room temperature in the tin, then move it to a cold place and keep it in the tin for at least another two hours, or until you're ready to serve it.

Fluffy gluten-free pancakes

This recipe makes the fluffiest pancakes, nutrient dense from homemade quark or chévre, fresh eggs, and goats milk. They are very filling and can be made either as sweet pancakes served as a treat with butter (by using the higher amount of honey in the recipe), or they can be made without any honey and served as a savoury pancake with chévre, smoked fish and other toppings, or as a side dish to borscht. I like to make these as smaller 'pikelet' type pancakes.

To get the most nutrition possible from the flour, you can leave the batter to soak for a few hours before cooking the pancakes.

This recipe makes enough for a hearty breakfast for two, or as a snack for a family. I usually double it to serve as a family meal. Leftover pancakes can easily be stored for later.

Ingredients
1 1/4 cups hulled buckwheat flour (or oat flour)
1/4 cup coconut flour
1/2 teaspoon bicarb soda
A pinch of salt
Optional pinch of vanilla powder, or 1 teaspoon vanilla extract
2 eggs
Optional 1-2 tablespoons honey
1/2 cup quark, ricotta, yoghurt or chévre
1/2 cup milk
1 1/2 teaspoons cider vinegar
Extra milk, if needed
Butter, for cooking with

Method

Combine the flours, bicarb and salt in a large mixing bowl. Make a well in the centre.

Whisk the eggs in a separate bowl, mix through the honey and quark until smooth, then add the milk, a little at a time, until it's evenly mixed. Stir through the cider vinegar, then pour this mixture into the bowl of flours and mix to form a thick batter. The batter should be very thick, but not like dough; add a little extra milk or water if you need to, or wait until you've cooked the first ones up to see how the batter behaves before adding any extra liquid.

Put one or two pancake pans on medium heat. Melt some butter in the pans and make sure the base is fully coated, then add a small amount of batter to each one. The texture of the first pancake from each pan will be a bit funny, so feel free to feed them to some friendly goats. Once you've cooked that first small pancake, either cook small pikelets or large pancakes until you've used all the batter up.

The batter is very thick, so you may need to spread it out in the pan slightly using a second spoon until it's around 1cm (1/2") high.

The temperature and cooking time will depend on your stove and pans, so it's hard to describe exactly. On electric stoves I usually start out at a medium high temperature and then turn it down to medium once the pan is hot.

Watch your pancake - if your cooking temperature is correct, you should notice the top of the pancake becoming less sticky with bubbles appearing. Flip it over once it isn't sticky, and then cook the other side for around 30 seconds, or until lightly browned.

Index

Cheese index

CPSIA information can be obtained
at www.ICGtesting.com
Printed in the USA
BVHW031106300819
557243BV00008B/101/P